VOLUME THIRTEEN

Advances in Planar Lipid Bilayers and Liposomes

EDITORIAL BOARD

VOLUME THIRTEEN

ADVANCES IN PLANAR LIPID BILAYERS AND LIPOSOMES

Editor

PROFESSOR DR. ALEŠ IGLIČ

Laboratory of Biophysics, Faculty of Electrical Engineering, University of Ljubljana, Ljubljana, Slovenia

Founding Editor

PROFESSOR DR. H.T. TIEN

Department of Physiology, Michigan State University, East Lansing, Michigan, USA

PROFESSOR DR. A. LEITMANNOVA LIU

Department of Physiology, Michigan State University, East Lansing, Michigan, USA

ELSEVIER

Amsterdam • Boston • Heidelberg • London • New York • Oxford
Paris • San Diego • San Francisco • Singapore • Sydney • Tokyo
Academic Press is an imprint of Elsevier

Academic Press is an imprint of Elsevier
Linacre House, Jordan Hill, Oxford OX2 8DP, UK
32 Jamestown Road, London NW1 7BY, UK
225 Wyman Street, Waltham, MA 02451, USA
525 B Street, Suite 1900, San Diego, California 92101-4495, USA

First edition 2011

Library of Congress Cataloging-in-Publication Data
A catalog record for this book is available from the Library of Congress

British Library Cataloguing-in-Publication Data
A catalogue record for this book is available from the British Library

ISBN: 978-0-12-387721-5
ISSN: 1554-4516

For information on all Academic Press publications visit our website at www.elsevierdirect.com

Transferred to Digital Printing in 2011

Contents

PREFACE

Volume 13 involves the chapters describing the mechanical properties of lipid bilayer membranes and protein–lipid interactions, lipid and membrane dynamics in biological tissues, membrane skeleton reorganization, photovoltaic solar energy conversion in biomembranes, multiparametric fluorescence approach in biomembrane experimental studies, and interactions between osteoblasts and titanium surface of implants. I would like to express my gratitude to all authors contributing their chapters, that is, to Profs. and/or Drs. S. Yoshida, K. Koike, T. Hianik, D. Kabaso, R. Shlomovitz, T. Auth, V. L. Lew, N. S. Gov, F. T. Hong, A. A. Heikal, E. Gongadze, Š. Perutková, V. Kralj-Iglič, and U. van Rienen. I also very much appreciate the continuous support of Ben Davie from Elsevier Office in London together with its coworkers from Elsevier's Chennai Office in India, Paul Prasad Chandramohan, Sunita Sundararajan, and Vijayaraj Purush. I would like to use this occasion to express my gratitude to the following members of the Editorial Board of APLBL: Prof. Sylvio May, Prof. P. B. Sunil Kumar, Prof. Nir S. Gov, Prof. Tibor Hianik, and Dr. Michael Rappolt. Special thanks to the previous editor Prof. Angelica Leitmannova Liu.

Aleš Iglič
Editor

CONTRIBUTORS

Thorsten Auth
Institute for Solid State Research, Research Centre Jülich, Jülich, Germany

Ulrich Beck
Institute for Electronic Appliances and Circuits, University of Rostock, Germany

Ekaterina Gongadze
Institute of General Electrical Engineering, University of Rostock, Rostock, Germany

Nir S. Gov
Department of Chemical Physics, Weizmann Institute of Science, Rehovot, Israel

Ahmed A. Heikal
The Department of Chemistry and Biochemistry, Swenson College of Science and Engineering, and The Department of Pharmacy Practice and Pharmaceutical Sciences, College of Pharmacy, The University of Minnesota Duluth, Duluth, Minnesota, USA

Tibor Hianik
Department of Nuclear Physics and Biophysics, Faculty of Mathematics, Physics and Informatics, Comenius University, Bratislava, Slovakia

Felix T. Hong
Department of Physiology, Wayne State University, Detroit, Michigan, USA

Aleš Iglič
Laboratory of Biophysics, Faculty of Electrical Engineering, University of Ljubljana, Ljubljana, Slovenia

Doron Kabaso
Department of Chemical Physics, Weizmann Institute of Science, Rehovot, Israel, and Laboratory of Biophysics, Faculty of Electrical Engineering, University of Ljubljana, Ljubljana, Slovenia

Kenzo Koike
Beauty Research Center, Kao Corporation, Tokyo, Japan

Veronika Kralj-Iglič
Laboratory of Clinical Biophysics, Faculty of Medicine, University of Ljubljana, Ljubljana, Slovenia

Virgilio L. Lew
Department of Physiology, Development and Neuroscience, University of Cambridge, Cambridge, United Kingdom

Šarka Perutková
Laboratory of Biophysics, Faculty of Electrical Engineering, University of Ljubljana, Ljubljana, Slovenia

Roie Shlomovitz
Department of Chemical Physics, Weizmann Institute of Science, Rehovot, Israel

Ursula van Rienen
Institute of General Electrical Engineering, University of Rostock, Rostock, Germany

Satoshi Yoshida
Gifu University, Gifu, Japan

CHAPTER ONE

Lipid and Membrane Dynamics in Biological Tissues—Infrared Spectroscopic Studies

Satoshi Yoshida[1,*] *and* Kenzo Koike[2]

Contents

Abstract

Functions of cell bilayer membranes are closely linked to dynamics and behavior of network of membrane components including lipids, proteins, and glycans. It is important to investigate the role of membrane components in the membrane functions without damage of the network of components. This is the reason why noninvasive and nondestructive analyses are so important for the study of the

* Corresponding author: Tel./Fax: +81 58 293 2655.
E-mail address: xyosida@gifu-u.ac.jp

[1] Gifu University, Gifu, Japan
[2] Beauty Research Center, Kao Corporation, Tokyo, Japan

Advances in Planar Lipid Bilayers and Liposomes, Volume 13
ISSN 1554-4516, DOI: 10.1016/B978-0-12-387721-5.00001-8

intact membranes and tissues. Vibrational spectroscopies including near- and mid-infrared absorption and resonance Raman scattering spectroscopies are useful for these purposes. In this chapter, we summarize the application of infrared spectroscopy to studies of lipids and bilayer membrane dynamics in biological tissues, and to the research for diagnosis of human diseases.

Abbreviations

18-MEA	18-methyl-eicosanoic acid
AGEs	advanced glycation end products
ATR	attenuated total reflectance
CERs	ceramides
CHOL	cholesterol
DEPE	1,2-dielaidoyl-*sn*-glycero-3-phospho-ethanolamine
DHA	docosahexaenoic acid
DMPA	1,2-dimyristoyl-*sn*-glycero-3-phosphate sodium salt
DPPC	dipalmitoylphosphatidylcholine
ESG	esterified sterylglucoside
ESI-MS	electrospray ionization-mass spectrometry
FFA	free fatty acid
FTIR	Fourier-transform infrared
GAGs	glycosaminoglycans
GalCer	galactocerebroside
GlcNAc	*N*-acetylglucosamine
HDL	high-density lipoprotein
IR	infrared
LDL	low-density lipoprotein
MALDI-TOF-MS	matrix-assisted laser desorption ionization time-of-flight mass spectrometry
NMF	natural moisturizer factor
NMR	nuclear magnetic resonance
PCA	principal component analysis
PLS	partial least square
POPC	palmitoleic phosphatidylcholine
SC	stratum corneum

1. Introduction

Cell membrane and tissue lipid dynamics are playing important roles in the life of all organisms on the earth. Especially, the cell membranes with phospholipid bilayer have mainly two functions (1) clear separation of the inner cell space from the outer world and (2) mediation of signal transduction and communications between the inner and outer spaces. These important functions of cell membranes are based on various biomolecular components, such as lipids (including fatty acids), proteins (enzymes, receptors, transporters, channels, and cytoskeletons), and glycans, and their interactions.

Functions of cell membranes are closely linked with dynamics and behavior of network of membrane components, such as lipids, proteins, and glycans with the aid of minerals and other nutrients. Thus, it is important to investigate the role of membrane components in the membrane functions without disruption of membrane integrity and without damage of the network of components. This is the reason why noninvasive and nondestructive analyses are so important for the study of membrane components in the intact membranes.

For noninvasive and nondestructive analyses of membrane systems, vibrational spectroscopies including near- and mid-infrared absorption and resonance Raman scattering spectroscopies are useful for these purposes. Vibrational spectroscopy can reveal the characteristics and interaction with the environment of biomolecules. Basically, the mid-infrared spectroscopy is based on the net changes of dipole moment of the molecules ("IR-active"), whereas the Raman spectroscopy is based on the change of electric polarizability of the molecules ("Raman-active"), and thus these spectroscopies are complementary techniques for studies in chemistry.

The mid-infrared absorption and Raman scattering spectroscopies are frequently used to study the interaction between lipids and proteins or glycans in the cell membrane, and this is due to the evidence that the energy level of hydrogen bonding or some interactions is comparable to the infrared energy level, and the interaction change between functional groups of membrane biomolecules (e.g., receptors, channels, glycoproteins, and glycolipids) and the surrounding water or lipid or protein molecules may be detectable in the infrared region.

Vibrational spectroscopy can reveal not only the characteristics of specific functional group, but also the mixture of many functional groups. As a matter of fact, the change of membrane dynamics may be in part the result of change of biomolecular networks in the membrane. This is because the application of statistical multivariate analysis or chemometrics to vibrational spectroscopy is useful and important for the noninvasive analyses of cell membranes, food staffs, and human tissues.

In the multivariate analysis or chemometrics for vibrational spectroscopy, the principal component analysis (PCA) and partial least square (PLS) regression analysis methods are frequently used for classification or categorization of phenomena, such as membrane dysfunction, or predication of change of factors which are, for example, some disease-related lipids and fatty acids. These statistical methods may provide vibrational spectroscopy—the chance to supersede the diagnosis of diseases with invasive clinical tests and destructive tests of food staffs.

The vibrational spectroscopy has been widely applied to the diagnostic analysis of diseases and nondestructive analysis of food staffs; however, at present there have been few vibrational spectroscopic techniques used as golden standard methods for diagnostic analyses. Usually biochemical analyses of blood or other body fluids have been used for the diagnosis as standard methods, and only one or a few specific factors could be measured in one shot of detection. For example, total cholesterol (CHOL), LDL, HDL, and triglycerides were mainly measured for diagnosis of hyperlipidemia which would be linked, in some cases, to complex metabolic diseases. Some complex diseases such as diabetic diseases and atherosclerosis may have complex pathogenic mechanisms with many onset factors and thus many biomarkers.

To realize those complex diseases and diagnosis, the technique which would detect many factors in a short time noninvasively by one shot of measurement would be very useful, and for this purpose the vibrational spectroscopy—especially mid-infrared spectroscopy—would be suitable because it can detect the change of properties of cell bilayer membranes including proteins, lipids, and glycans simultaneously, and actually the near- and mid-infrared spectroscopy as well as Raman scattering spectroscopy may have a great potential in the clinical diagnostics field.

In this chapter, we summarize the advances of infrared spectroscopic techniques in the study of cell membranes and lipids and their interactions with other biomolecules in relation to the diagnosis of human diseases.

2. Historical View of Vibrational Spectroscopic Studies on Lipids and Membrane Dynamics in Biological Tissues

Application of vibrational spectroscopy, especially mid-infrared spectroscopy, to the studies of cell membrane and lipids has a long history and has provided a lot of data concerning to the study of membrane and lipid dynamics in human tissues.

2.1. Membrane Fluidity

Infrared spectroscopy could reveal the fluidity change of the cell membranes and liposomes with measuring the methylene CH stretching mode of the membrane fatty acyl chains [1,2]. Membrane fluidity could be measured also by other methods, such as using electron spin resonance and fluorescence probes [3], and the viscosity change corresponding to the capability of moving of fatty acyl chains in the membrane was detected as a kinetic parameter of microenvironment of the probes. On the other hand, the measurement of membrane fluidity by infrared spectroscopy may show the extent of packing of membrane acyl chains, or in another words "softness of the membrane" which is a static parameter of macroenvironment.

Actually, the change of membrane fluidity may be measured by the change of infrared absorption peak position for methylene CH symmetric stretching mode at around 2850 cm^{-1} [1,2] as well as NMR measurement [4]. In the fluidity measurement, the peak shift to lower wave number direction indicated that the membrane became harder or ordered in the lipid network structure, whereas the shift to higher wave number direction indicated softer or disordered [5]. The infrared spectroscopic analysis could be used for elucidation of phase behavior of long-chain phospholipid membranes [2].

In lipid bilayer membranes, acyl chains in the lipids are interacted with each other by the van der Waals or London forces. When the interaction among methylenes was increased with the increase of packing of the membrane acyl chains which were attracted with each other by London forces, methylene CH bond force constant might be decreased. This would shift the CH symmetric stretching mode-originated infrared absorption peak to the lower wave number (energy) direction, for example, a shift from 2853 to 2852 cm^{-1}, indicating hardening of the membrane and usually observed when the temperature was decreased from 37 to 10 °C for pig brain microsomal membranes (unpublished result).

In another case, the Fourier-transform infrared (FTIR) spectroscopic analysis of lipid O=P=O, C=O, and C–H vibrational bands of POPC/CHOL (palmitoleic phosphatidylcholine/cholesterol mixture) liposomes revealed an increase in the conformational order of the acyl chains at or close to the predicted critical cholesterol molar fractions [3]. Here, the shift of methylene infrared absorption peak was observed from 2851.2 to 2850.7 cm^{-1} when the fraction of cholesterol was increased from 0% to 40%. This indicated that the insertion of cholesterol to POPC model membrane contributed to the increase of packing of membrane acyl chains.

For measurement of membrane fluidity in the biological tissues such as artery, it would be difficult to use probe methods (using fluorescence anisotropy measurement and electron spin resonance spectroscopy) because the probe methods would require the insertion of external probe to the tissue membranes and the suitable insertion of probe might be practically impossible to the arterial membrane. Infrared spectroscopy provides a technique that does not require the insertion of probes, and this is especially advantageous for measurement of biological tissues.

Actually, the application of FTIR for measurement of mouse pulmonary artery was reported [6] and the change of conformational disorder (fluidity) in the artery membrane lipids could be detected. In this case, the arterial tissue was sandwiched by CaF_2 disks and the transmission FTIR absorption spectrum could be measured. In this report, the average conformational disorder in membrane lipids in the pulmonary artery *in situ* was increased by the treatment of monocrotaline which was a toxic alkaloid and caused early pulmonary endothelial injury with gradual development of pulmonary hypertension. Moreover, the membrane fluidity of carotid tissue of spontaneously hypertensive rat could also be measured *in situ* [7], and the membrane fluidity was increased in the hypertensive rat carotid subjected to anoxia, but not in the control rat. In the carotid or pulmonary arterial tissues, the increase of membrane conformational disorder would be linked to vulnerability of the tissue membrane.

2.2. Hydration of Membrane

The state of hydration of membrane surface may affect the hydrolytic activity in or on the surface of membrane, and the hydration may be closely related to the presence of glycans on the membrane, especially glycosylated lipids and proteins. Actually, it was reported that the interfacial properties of cell membranes were important and phospholipase A_2 activity was influenced by hydration (or in another word, water activity) of the membrane surface [8–10], and thus the inflammation was affected.

Hydration state around carboxyl and phosphate groups may be affected by the presence of polyhydroxy compounds (such as glycans) or by changing the chemical groups esterified to the phosphates, mainly choline, ethanolamine, or glycerol. Thus, surface membrane properties, such as the dipole potential and the surface pressure, are modulated by the water at the interphase region by changing the structure of the membrane components [10–13].

The hydration or dehydration of membrane surface could be measured by FTIR using the absorption of hydroxy (–OH) and carboxylate (–COO–) groups which would be changed by the change of hydrogen bonding with water around the residues on the membrane surface. It was reported previously [8] that nondestructive FTIR analysis could detect the modification of rat brain microsomal membranes and these modifications of brain microsomal membranes were dependent on the dietary fatty acids and learning behavior. In this report [8], FTIR spectral differences for brain microsomes were observed mainly in the absorption bands of fatty acyl ester at around 1730 cm^{-1} (*sn*-2 position), phosphate ester and oligosaccharides in the range of 1050–1100 cm^{-1}. The infrared band of fatty acyl ester at the *sn*-2 position in the microsomal membrane shifted to a higher wave number position (1731 cm^{-1}) in the perilla oil-diet group (α-linolenic acid-rich) than that in the safflower oil group (α-linolenic acid-deficient) at 1727 cm^{-1} after the learning behavior, suggesting a difference between both groups in hydrogen bonding of the fatty acyl ester with water. Without learning behavior, both groups showed similar ester absorption at *sn*-2 position (1729 and 1728 cm^{-1}). The infrared band shift observed for the *sn*-2 ester was interpreted to be mainly due to the change of hydrogen bonding strength between the *sn*-2 fatty ester (C=O) and the surrounding water molecules.

2.3. Microdomain Structure of Membrane

It has been reported that the normal cell membranes may contain microdomains which may play various roles in cell functions such as intracellular trafficking, signal transduction, and cell–cell recognition [14,15], and the FTIR spectroscopy could show the presence of microdomains in cell membranes [16]. This microdomain has been named sometimes as "raft" or "nanoclusters" [17,18].

A report [16] showed that the membrane assemblies in the liquid crystalline phase were prepared and then the membrane dispersions were rapidly cooled to induce the gel phase where lateral diffusion was dramatically inhibited, and the lipid microdomains were found by FTIR and ultrasonic velocimetry to be in a nanometer size in scale, in contrast to the micron-sized domains more commonly found in phase-separated systems [19,20]. Here, methylene (in perdeuterated DPPC-d_{62}; dipalmitoylphosphatidylcholine) infrared absorption at around 1090 cm^{-1} for CD_2 methylene deformation (bending) band was split at $-120\ °C$ with a splitting of $8.1 \pm 0.2\ cm^{-1}$, and the CH_2 deformation (bending) mode at around 1470 cm^{-1} of GalCer (galactocerebroside) and DPPC mixture also showed splitting at $-120\ °C$.

In this report, the CH_2 deformation mode existed as a single peak at 1469 cm^{-1} at 20 °C but was split in a manner similar to that of the pure GalCer dispersion at $-120\ °C$. Two separate splitting parameters in the second derivative spectrum were observed. The inner splitting, $2.2 \pm 0.8\ cm^{-1}$, corresponded to the sphingosine chain, whereas the outer splitting of the stearoyl chain was $7.0 \pm 0.1\ cm^{-1}$. This outer splitting for GalCer in the DPPC-d_{62} bilayer matrix corresponded to approximately six chains. DPPC-d_{62} also evinced a temperature-dependent change, representative of microdomain formation. The microdomains in the membrane may have different fluidity or viscosity from the other region of the membrane, and thus the methylene infrared absorption band would be different.

2.4. Protein–Lipid Interactions

Protein–lipid interaction may play a key role in the function of cell membranes, and several methods were applied to detect the interactions, and the FTIR was one of those methods [21,22].

A report [23] showed that FTIR spectroscopy indicated the presence of interaction of P_γ (heterotrimeric G protein γ-subunit peptide) and P_γ-FN (the farnesylated peptide) with the polar and interfacial regions of phospholipid bilayers. The binding of P_γ-FN to model membranes was due to the farnesyl group and positively charged amino acids near this lipid. On the other hand, membrane lipids partially altered P_γ-FN structure, in turn increasing the fluidity of phospholipid membranes. These data highlight the relevance of the interaction of the C-terminal region of the G_γ (G protein γ-subunit) protein with the cell membrane and its effect on membrane structure. In this report, 1,2-dielaidoyl-*sn*-glycero-3-phospho-ethanolamine (DEPE) and 1,2-dimyristoyl-*sn*-glycero-3-phosphate sodium salt (DMPA) were used as the membrane lipids.

In this report, P_γ-FN and P_γ exhibited an amide I band characteristic of β-sheet structures with maxima at 1625 (P_γ-FN) or 1638 (P_γ) cm^{-1} and a weak component at around 1688 cm^{-1}. Amide I spectra of both peptides also displayed a random-coil component with a shoulder at around 1645–1649 cm^{-1}. A centrifugation assay was used to analyze the membrane binding of both peptides qualitatively under the conditions used in FTIR experiments. The band at 1625 cm^{-1} was more prominent in the presence of DEPE and even more so in the presence of DMPA, indicating an increase in the proportion of P_γ-FN that adopts a β-sheet structure as a result of its association with DEPE or DMPA lipids. In the case of the P_γ peptide and in the presence of DMPA membranes, the amide I band was observed in the membrane pellet. The spectra of P_γ in DEPE membranes showed two main bands at 1652 and 1633 cm^{-1}, which clearly differed from the FTIR spectrum of P_γ alone. This might indicate that the association of the peptide with DEPE bilayers induced a partial loss of the β-sheet secondary structure, adopting α-helical and/or random-coil structures.

As shown in this report typically, the membrane lipids may affect the secondary structure of membrane proteins and peptides, and this structural change may cause functional changes of the proteins [24,25]. On the other hand, *vise versa*, the membrane structures may be changed by the presence of peptides and hydrophobic proteins [26,27]. These changes of protein–lipid interactions were well detected by FTIR when the experiment was suitably designed.

3. Membrane Lipids and Fatty Acids in Human Tissues

For human tissues, the measurements of proteins, glycans, lipids, and fatty acids by infrared spectroscopy were frequently performed to investigate the link between the diseases and the changes of tissue/cell compositions and characters [28–30].

Diseases of human—such as cancer, diabetes, and cardiovascular diseases—may have various aspects of tissue/cell structural and functional changes, and these diseases may accompany many molecular and signal network changes. Some cancers may include an increase in the nuclei-to-cytoplasm ratio, increase of the relative amount of DNA, enhancement of the phosphorylation of proteins, a decrease in the glycogen level, a loss of hydrogen bonding of the C–OH groups in the amino acid residues of proteins, a decrease in the overall CH_3-to-CH_2 ratio, and an accumulation of triglycerides [28]. The most significant merit in using FTIR for measurement or diagnosis of these tissue/cell changes is that FTIR may be able to detect these changes of biomolecular functional groups simultaneously and noninvasively by one shot of measurement.

3.1. Cancer Tissues

Initially, the application of FTIR to the study of cancers was performed by Rigas's group [28] in 1990, and human colorectal cancers displayed abnormal FTIR spectra. It was reported that all cancer specimens displayed abnormal spectra compared with the corresponding normal tissues, involving the changes in the phosphate and CO stretching bands, the CH stretch region, and the pressure dependence of the CH_2 bending and C=O stretching modes. It may be interesting to note that in malignant colonic tissue, the intensity of methyl (CH_3) band at 2958.5 cm^{-1} was decreased while that of the symmetric methylene (CH_2) band at 2852.5 cm^{-1} was increased, when compared with the corresponding bands in normal tissue. This indicated that the ratio of the number of methyl groups to that of methylene groups was decreased in malignant tissues, compared with normal colonic tissue, indicating that the lipids or fatty acids were relatively increased in the colonic cancer tissue. In this report, the pressure dependence of the CH_2 bending mode at around 1470 cm^{-1} was also measured to investigate the packing characteristics of methylene chains in the membrane lipids. At atmospheric pressure, the frequency of this band was higher in the malignant tissue than that in its normal counterpart, and this relationship was reversed as pressure increased.

Normally, colorectal cancer comes in many forms and main form is adenocarcinoma. Rigas and Wong [29] reported in another paper that the pressure dependence of the frequency of the CH_2 bending mode at around 1470 cm^{-1} had been used to study interchain packing and order/disorder properties of lipid bilayers, and the pressure dependence of the mode for HCT15 adenocarcinoma cell line was very close to that displayed by normal colon tissues. This indicated one of two possibilities that either this property was not critical to malignant transformation or the cultured colonocytes had dispensed with such a requirement.

Recently, the FTIR spectroscopic studies for cancer cells of esophagus [31] and viral cancer progression of cervical tissues [32] were reported.

Maziak and Wong's group [31] reported in 2007 that specific changes were observed in the FTIR spectral features of esophageal cancer, including a *decrease* in the overall CH_3-to-CH_2 ratio and an accumulation of triglycerides, an enhancement in the phosphorylation of proteins, and so on. On the other hand, Bogomolny and Huleihel's group [32] pointed out that the infection of murine sarcoma virus to murine fibroblast cell lines (NIH/3T3) and mouse embryonic fibroblast (MEF) cells progressed malignancy in those cells and cervical tissues and the protein/lipid ratios (CH_3-to-CH_2 ratio) were *increased* in the progressed cancer cells and tissues, and this is a contradictory result against the Rigas's group data [29,30] where CH_3-to-CH_2 ratios were decreased in the colon and esophageal cancer cells. These results may indicate that the change of lipid metabolism in cancer cells, that is, whether the accumulation of triglyceride occurred or not, may be dependent on the cancer cell types.

3.2. Fatty Acid Metabolism and Metabolic Diseases

Epidemiological research using fatty acid compositional analyses of human body fluids (blood, etc.) and tissues may be important for assessment of the effectiveness of dietary fatty acids on the physiology of the human body. However, the measurement of fatty acid compositions in tissues involves tedious and time-consuming procedures for extraction and derivatization (methyl ester, etc.) of fatty acids from tissues for subsequent analyses with gas chromatography or high-performance liquid chromatography. These standard chromatographic methods may not be practically suitable for a large-scale epidemiological research. On the other hand, vibrational spectroscopies such as Raman and infrared spectroscopies may be useful techniques to measure lipid compositions in food staffs and human tissues and fluids. It has been reported that nondestructive FTIR–ATR measurement of human oral mucosa [33] could monitor diurnal changes of polyunsaturated fatty acids in the mucosa. On the other hand, Lam *et al.* [34] also reported that FTIR and chemometrics revealed low-density lipoprotein oxidation with protein conformational changes and provided a simple rapid technique for measuring primary and secondary oxidation products.

Furthermore, the fatty acid compositions of human oral mucosa could be measured using FTIR–ATR method [33,35]. In this report, the second derivative infrared spectra of the mucosal tissues in the wave number regions from 1600 to 1760 cm^{-1} and 2800 to 3050 cm^{-1} were analyzed with PLS multivariate regression analysis method. The predicted values were compared with the measured values of 10 categorized fatty acid compositions, that is, a (saturated C17 or lower), b(C16:1 + C17:1), c(C18:0), d(C18:1), e(C18:2), f(saturated C20 or longer), g(C20:3 + C20:4), h(C22:1 + C24:1), i(C22:6), and j(γC18:3), where γC18:3 is γ-linolenic acid. Almost all fatty acid compositions of oral mucosa were well predicted with difference between predicted and measured values within ±5% of total; however, errors were relatively larger in minor components such as C22:6 than major components.

Many papers have been published concerning multivariate analysis methods for fatty acid compositional analyses [36–38] using vibrational spectroscopy. Beattie *et al.* [36] reported that the bulk unsaturation parameters and abundance of polyunsaturated fatty acids of several adipose tissues could be predicted by Raman spectroscopy using multivariate analysis. They reported that Raman analysis has the advantage of giving good correlation coefficients without the need for prior solvent extraction steps in the analysis. On the other hand, Ripoche and Guillard [37] reported when using FTIR that the prediction of fatty acid composition of unextracted adipose tissue gave poor correlation coefficient ($R^2 = 0.69$–0.79). Improvement of the coefficient in the reported case [33] may result mainly from using less noisy second derivative infrared spectra in the selected wave number regions. Afseth *et al.* [38] reported using Raman and near-IR spectroscopic methods that Raman spectroscopy was suitable for predicting the total level of unsaturation, and both Raman and near-IR spectroscopy were feasible for rapid quantification of fat composition in complex food model systems.

Moreover, it was reported that FTIR–ATR technique could monitor the metabolism of docosahexaenoic acid in the adult body *in situ* and noninvasively [39]. Normally, the ingested lipids in foods were digested and adsorbed through intestine to blood and to liver, and resorbed through vessels to peripheral tissues, including skin tissues. It may be essential to find out sites of human body, especially skin tissues, where the lipid and fatty acid changes of blood were reflected nearly immediately, to measure noninvasively with FTIR–ATR. In this chapter, the change of human lip surface lipid compositions reflected well that of ingested lipids, and the changes of lipids and fatty acids were really detected with FTIR–ATR *in situ*. Human lip has generally four portions anatomically (1) appendage-bearing epidermis, (2) keratinized vermilion (red zone), (3) parakeratinized intermediate zone, and (4) labial mucosal epithelium. Yoshida *et al.* [39] measured the lower lip, that is, keratinized vermilion surface (outermost stratified corneum) by FTIR–ATR method, and this portion was suitable to measure the metabolism of polyunsaturated fatty acids originated from diet.

Diabetes mellitus is one of the metabolic diseases, and normally the diagnosis of diabetes is usually done by using the measurement of blood glucose and hemoglobin A1c, a glycosylated protein. Recently, it was reported [40] that the FTIR spectral analysis revealed differences in several major metabolic components—lipids, proteins, glucose, thiocyanate, and carboxylate—that clearly demarcated healthy and diseased saliva. In this report, differences in Fourier self-deconvolution processed mean IR spectra were noted from normal and diabetic saliva, and the difference spectrum helped identify molecular components most at variance between two groups of spectra. For example, the lipid ester band at 1735 cm^{-1} was more intense in the diabetic saliva, while the bands located at 1400 and 1582 cm^{-1}—the symmetric and asymmetric carboxyl radical stretching vibrations of carboxylate groups such as those in lactate or side chains of protein in saliva—were decreased for diabetic saliva. The role of lactate in the modulation of hormone release and responsiveness and in the control of homeostasis was recently reviewed by Sola-Penna [41]. Another spectral area including 950–1180 cm^{-1} for C–C/C–O stretch of sugar moieties and 1020 cm^{-1} for C–O of glycogen was contributed from AGEs (advanced glycation end products) of diabetic saliva.

Even very recently, a number of reports have appeared concerning to the application of infrared spectroscopy to measurement of biomolecules including lipids of human tissues and cells nondestructively [42–51], and these reports may indicate that the infrared spectroscopy or vibrational spectroscopy has been increasingly recognized to be useful for clinical diagnosis of human diseases.

4. Skin and Hair Lipids and Membranes

The structure and function of lipids and membranes in skin surface tissues, that is, stratum corneum (SC) and hairs, were quite different from those in other normal organs, such as liver and heart tissues. These differences may be realized by the very special roles of skin SC and hairs, and both SC and hairs are integrated cells without living. Even though these tissues are not alive, they really play a protective and important role for the body, and the integrity of those "dead cells" is indispensable for the functions. Lipids and their interactions with proteins are especially important for the integrity of cells and matrix and for the barrier function of those tissues. In this section, we show data obtained mainly from our laboratory.

Figure 1 shows the FTIR–ATR spectra of human face skin (stratum corneum) in the two zones: U3 (the chin skin) and T1 (the forehead skin). This suggested that T1 skin SC contained more lipids than U3 skin SC. It was also noted that the fingerprint areas (1000–1400 cm^{-1}) in both zones were significantly different in the IR spectra.

Figure 2 shows the FTIR–ATR spectra of human hairs, outer and inner surfaces, measured by microscopic FTIR measurement system. The inner surface showed a clear peak at around 3010 cm^{-1}.

4.1. Protein-Bound Fatty Acid and Ceramides in Hairs

Human scalp hairs are comprised of proteins (>90% in dry weight) and lipids (<10%) and those components provide hair fibers with suitable elasticity and mechanical resistance against elongation and torsion. Hair lipids [52] are mainly composed of squalene, wax esters, triacylglycerol,

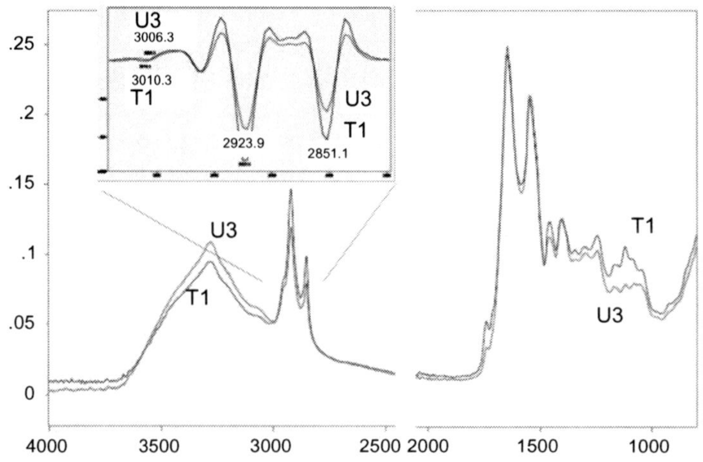

Figure 1 FTIR–ATR spectra of human face skin in T1 and U3 zones.

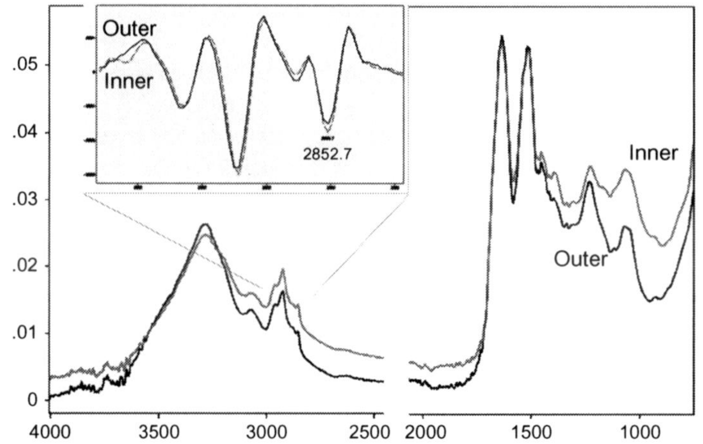

Figure 2 FTIR–ATR spectra of human hairs, outer and inner surfaces. Inserted are the second derivative spectra at around 3010 cm^{-1} and 3000–2800 cm^{-1} regions for CH stretching vibration of *cis*-unsaturated C–C bonds and methylene/methyl CH vibrations, respectively.

free fatty acids, cholesterol, ceramides, cholesterol sulfate, 18-methyl-eicosanoic acid (18-MEA), and other minor components.

Among these lipids, only 18-MEA binds covalently to the cuticle surface proteins with a thioester linkage. 18-MEA is known as a specially designed branched-chain fatty acid and makes the hair surface hydrophobic and acts as a boundary lubricant to decrease friction resistance with providing a large segmental volume of the anteiso-moiety which exhibit liquid phase-like behavior on the hair cuticle [53].

18-MEA could be detected normally by gas chromatography after removal from hair proteins with saponification. The measurement of hair cuticle surface by FTIR–ATR may detect 18-MEA as a major branched-chain fatty acid. As shown in Fig. 2, the FTIR spectra of hair outer and inner surfaces indicated that the methylene CH symmetric stretching mode was observed at 2852.7 cm^{-1}, which was a little higher wave number position than that of skin methylene mode observed at 2851.1 cm^{-1} (Fig. 1). Here, the inner surface was prepared by cutting a hair fiber in the middle with a microplane. Figure 2 also suggested that unsaturated fatty acids were much less in the outer surface of cuticle than the inner surface of hair because the outer surface did not show clear IR absorbance at around 3010 cm^{-1} (originated from CH stretch of *cis*-CH=CH double bond) whereas the inner surface had a clear absorbance at around 3010 cm^{-1}. These results indicated that the phase of 18-MEA bundles on the cuticle surface was in more fluid or disordered state than skin surface, and these data were consistent with the idea that the anteiso-moiety of 18-MEA exhibited liquid-like behavior. The relatively disordered state of 18-MEA may be resulted from the presence of 18-methyl branch in the fatty acyl chain, but not from the unsaturated bonds in acyl chains which were practically absent on the surface of hair cuticle.

Ceramides are contained in human hairs as well as skin, and especially nonhydroxy or α-hydroxy fatty acid moiety and a dihydrosphingosine moiety are thought to be present in hair CERs. It has been hypothesized that the CERs were related to apoptosis during keratinization in living hair matrix cells to dead cuticular or cortical cells and contributed to barrier function and water holding in hair. It was reported [54] that the composition of CERs in hair was characterized by predominant CERs with saturated or unsaturated and carbon even-numbered fatty acid moieties and C18 dihydrosphingosine moiety, and also by isomeric CERs, α-hydroxy fatty acid-containing CERs, and odd chain-containing CERs.

As shown in Fig. 2, the IR absorption intensity of amide II (at around 1550 cm^{-1}) or amine-derived band (at around 3300 cm^{-1}) or NH bending/stretching and C–N stretching mode was relatively higher in hair than in skin (Fig. 1), and this may indicate the mixing of IR spectra of amide-related band of various CERs which had various hydroxy forms and various peak positions at 1535–1560 cm^{-1} [55]. So far, there is no report which showed that the CERs in hair or skin were directly observed *in situ* by FTIR–ATR, but somehow the amide bands in FTIR of hair or skin may be affected by the presence of CERs.

4.2. Minor Lipids in Hair

As described earlier, major lipids in hair medulla are squalene, wax esters, triacylglycerols, free fatty acids, cholesterol, ceramides, and cholesterol sulfate. 18-MEA is present on the cuticle. We found also minor lipids present in the whole hair shaft, and also in hair bulb (root) as described below. Figure 3A shows the silica gel–TLC pattern of the hair extracted lipids (developed with chloroform:methanol (95:12) solvent mixture), and the minor component is circled on the figure and is called “lipid X” here. The “lipid X” extracted from the silica gel plate was dissolved in methanol and measured by FTIR shown in Fig. 3A (right). This IR spectrum clearly indicated the presence of amide bond in the lipid X compound, and also fatty acyl (at around 2850 cm^{-1}) and glycan (at around 1050 cm^{-1}) structures. The ESI-MS spectrum of this compound (lipid X) was compared with the standard esterified sterylglucoside (ESG) as shown in Fig. 3B because lipid X was developed very near to the position of ESG in the TLC (Fig. 3A, left). The ESI-MS patterns were similar with each other; however, mass peak positions were different. Especially, the widely distributed fine peaks cluster at around m/z 910 ($\pm$60) was found in the standard ESG with constant mass peak gas of Δ16 mu, and that at around m/z 840 ($\pm$70) was in the lipid X compound with various mass peak gaps of Δ12, Δ14, Δ24, and Δ26 mu. Further analysis of the compound was done after acidic hydrolysis and the materials in aqueous phase were observed with ESI-MS method as shown in Fig. 3C.

In Fig. 3C, ESI-MS spectra of the standard acetylated glucose and acetylated *N*-acetylglucosamine (GlcNAc) were presented (upper panel), and the mass pattern of acetylated materials of the hydrolyzed and extracted “lipid X” compound in aqueous phase was also presented (lower panel). When both acetylated GlcNAc and acetylated soluble “lipid X” fragments were compared, the ESI-MS patterns of both were almost identical. Acetylated glucose was also observed, and these results suggested that lipid X composed of GlcNAc and glucose as glycan structures. We observed that cholesterol and lanosterol were contained in the hydrophobic fraction of hydrolyzed lipid X compound (not shown in this section), and summarized the candidate structures of the ESG-like compound (lipid X) in Fig. 3D. All mass patterns could be reasonably explained by the combination of various fatty acids and glucose or GlcNAc, and cholesterol or lanosterol. As GlcNAc contains amide bond in the structure, this mass spectral result was consistent with the previous FTIR data which suggested the presence of amide bond in lipid X.

On the other hand, hair bulb obtained from hairs naturally removed was investigated by FTIR and mass analysis for the extracted lipids from the hair bulb. In Fig. 3E, a microphoto of a typical hair bulb was shown after cut off from the shaft. We could detect IR spectra at the hair bulb top (root side), middle, and the site near shaft separately by FTIR–ATR with microscope (IlluminatIR; SmithDetection, Inc., UAS). Normally, the IR spectrum of hair shaft surface was different from that of hair bulb or root especially in the IR absorption intensity at 1395 cm^{-1}, and this IR absorption was much higher in bulb than in hair shaft surface. Even within the bulb, the peak ratios between

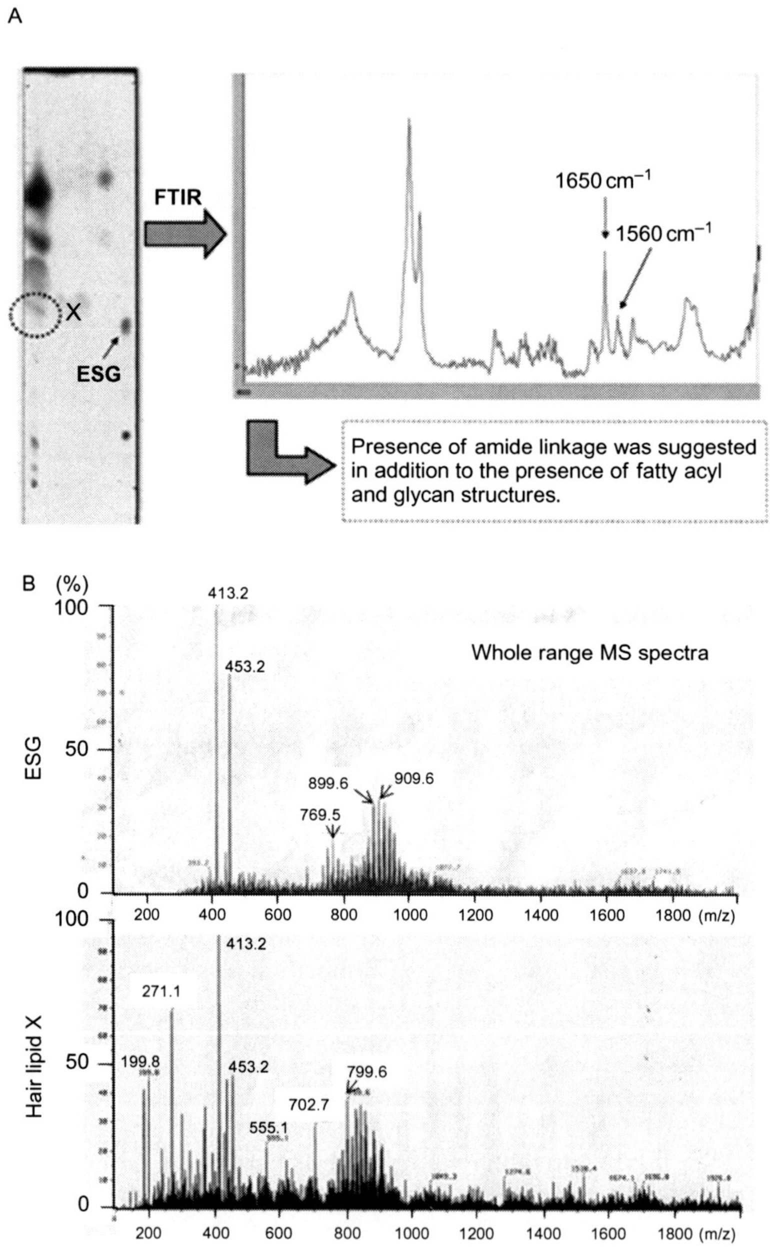

Figure 3 (Continued)

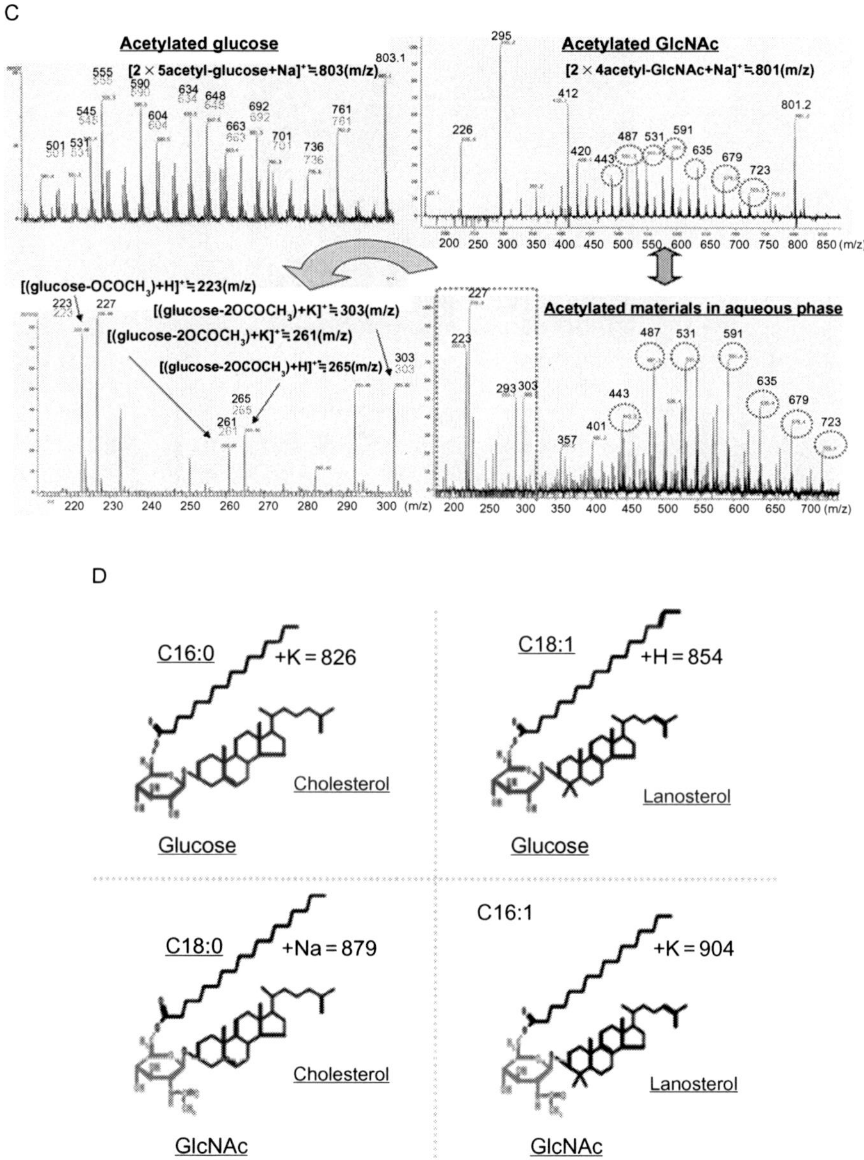

Figure 3 (Continued)

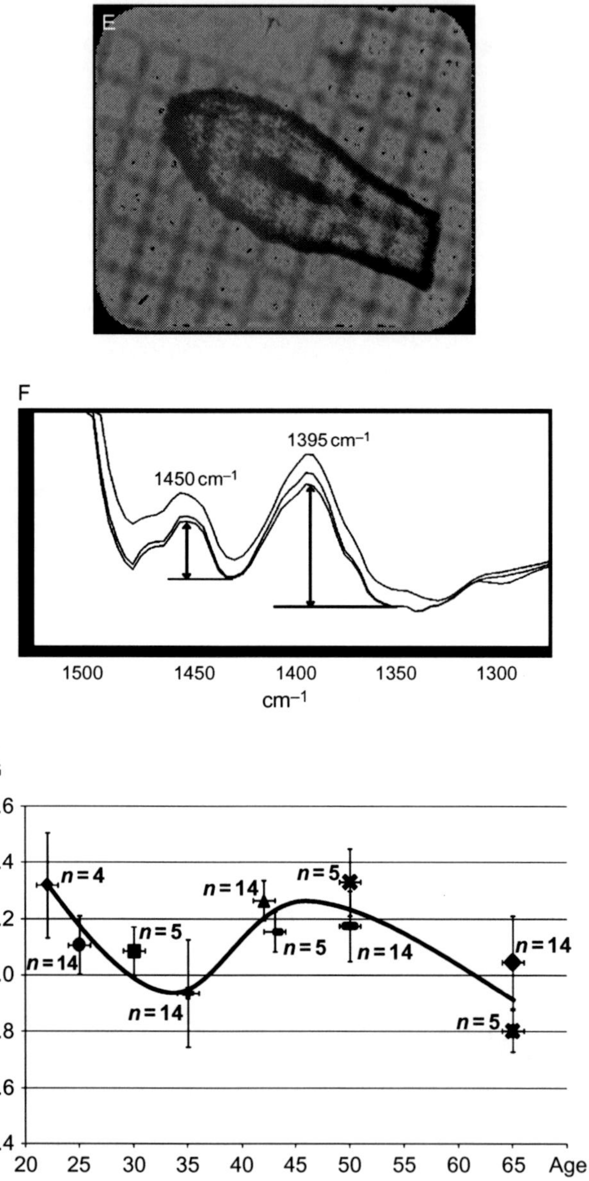

Figure 3 (Continued)

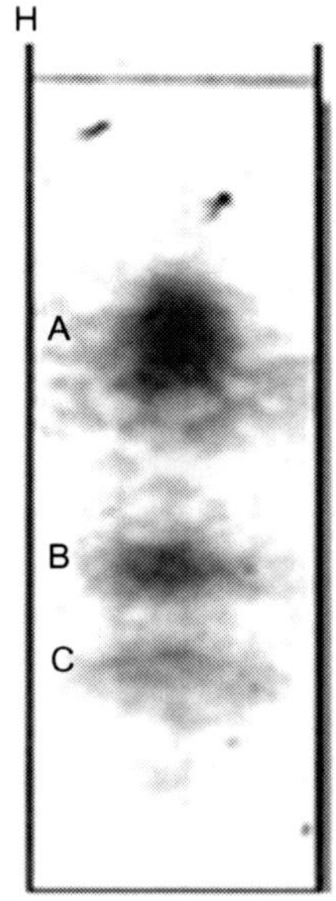

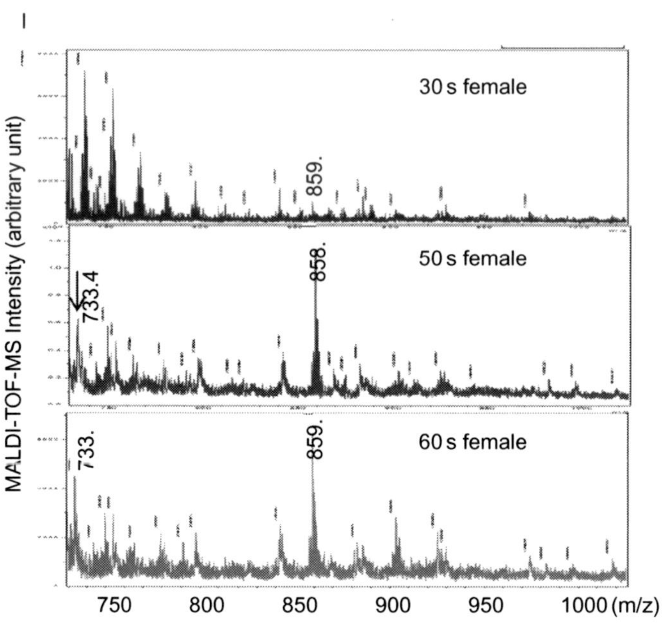

Figure 3 (A) Ester sterylglycoside-like lipids in human hair shaft detected by TLC and FTIR. Left: the silica gel–TLC patterns of the hair extracted lipids. Right: FTIR spectrum of "lipid X." (B) ESI–MS spectra of ester sterylglucoside (standard) and "lipid X." (C) ESI–MS spectra of saccharides in aqueous phase in ester

($1395\ cm^{-1}$) and ($1450\ cm^{-1}$) were different among these sites, and the ratio was higher at the bulb top (root side) than the site near shaft (Fig. 3F).

The age-dependent changes of the ratio at the middle of the bulb were shown in Fig. 3G. The subjects from whom the hairs were collected were all women from 20s to 60s of ages (*n* was the number of hair bulbs measured). This figure indicated that the ratio became smaller in 30s of age, and again increased in 40s and 50s, then decreased again in 60s. The hair bulb component which showed a typical IR absorption at $1395\ cm^{-1}$ may be regulated in the age-dependent manner, and probably by hormone.

We had revealed that the $1395\ cm^{-1}$ component in the hair bulb was a lipid-like material, and the typical TLC pattern of extracted lipids from the bulb was shown in Fig. 3H. This pattern was obtained in the silica gel plate with chloroform/methanol (95:12) as a development solvent which was used for hair shaft extract shown in Fig. 3A.

In Fig. 3H, the spot A contained main neutral lipids such as squalene and triglycerides and spot B may contain free cholesterol. The spot C was supposed to contain free fatty acids (FFAs) and ceramides (CERs) and other minor components, and ESG-like compound found in hair shaft may also be contained in spot C. The relative spot intensity of spot C as the ratio of spot C to spot A showed the age dependencies, and the intensity of spot C was the lowest for 30s of age and higher in 40s–60s of ages (not shown in this section).

The components of spot C were analyzed by MALDI-TOF-MS spectrometry (with 2,5-dihydroxybenzoic acid as a matrix) after extraction from the silica gel. We found mass peaks which showed the lowest for the spot C extract of 30s of age and higher for 40s–60s of ages, and those peaks were at *m/z* 858–859 as shown in Fig. 3I. Normally, in MALDI-TOF-MS spectrometry, free fatty acids (mainly C16–C18) and ceramides may show mass peaks below *m/z* 700 (typically *m/z* 500–650 for ceramides), and a peak at *m/z* 733 may represent a reduced form of C16:0 sphingomyelin. The mass peaks at *m/z* 858–859 were in the range typically observed for ESGs as shown in Fig. 3A–D.

However, fatty acyl ESGs had no strong IR absorption at around $1390\ cm^{-1}$, so it may be reasonable to assume that the hair bulb-specific component with IR absorption at $1395\ cm^{-1}$ might be the modified fatty acyl ESG, and it has been known that typical IR absorption at 1390–$1400\ cm^{-1}$ was originated from nitrated compound (modified by NO_2),

---

sterylglucoside-like lipids. Acetylated glucose and acetylated N-acetylglucosamine were used as standards. (D) Candidates of esterified sterol glycolipids in human hair shaft. (E) A typical microphotograph of a hair bulb after cutting off from the shaft. (F) FTIR spectra of hair shaft surface and bulb at around $1400\ cm^{-1}$. (G) Age-dependent changes of FTIR peak ratios between ($1395\ cm^{-1}$) and ($1450\ cm^{-1}$) at the middle of bulbs. (H) A typical TLC pattern of extracted lipids from the bulb. (I) Age-dependent MALDI-TOF-MS profiles for esterified sterylglucoside in spot C of (H).

and actually the synthesized nitro-lanosterol compound in our laboratory showed IR band at around 1390 cm^{-1} (not shown in this section).

Finally, we assumed that the compound in hair bulb (in spot C of Fig. 3H) which had IR absorption at 1395 cm^{-1} and the mass peak at around *m/z* 859 might be the nitrated compound with glucose + cholesterol + C18:1, which showed calculated molecular mass at *m/z* 859.2, as a candidate.

4.3. Effect of Dietary Lipids on the Hair Lipids

Hair lipids and fatty acids may be supplied originally from dietary lipids and thus from blood, and metabolized in hair follicle to synthesize hair-specific components, such as type I and II hair keratin proteins, 18-MEA fatty acid, and various lipids including ceramides and sterol glycosides. However, direct relationship has not been demonstrated between dietary lipid intake and the storage of lipids in hair fibers so far, and here we show a part of data suggesting the correlation between dietary fat intake and appearance of lipids in hair.

Figure 4 indicates the method to measure FTIR spectrum of one hair fiber with FTIR–ATR spectroscopy with microscope (IlluminatIR; Smith-Detection, Inc., USA), and the right inserted photo showed the scale (110 μm) and actual hair fiber. The measurement of half-cut hair fiber surface was done by attaching the ATR probe on the points marked in Fig. 4 (left) with 110 μm intervals.

In Fig. 5A, the scanning electron micrograph of half-cut hair was demonstrated, and many fine fibers were present. Figure 5B showed the presence of periodic changes of relative IR intensities at 2960 cm^{-1} in the direction along longitudinal axis of hair fiber for four subjects (two men and two women), and the fine periodicity (point-to-point period was nearly 8 h) was averaged to obtain one point as the 1-week-averaged data. Roughly, the length (0.33 mm) containing three points would correspond to nearly 24 h as the scalp hairs are thought to grow with normally 0.3–0.4 mm long a day.

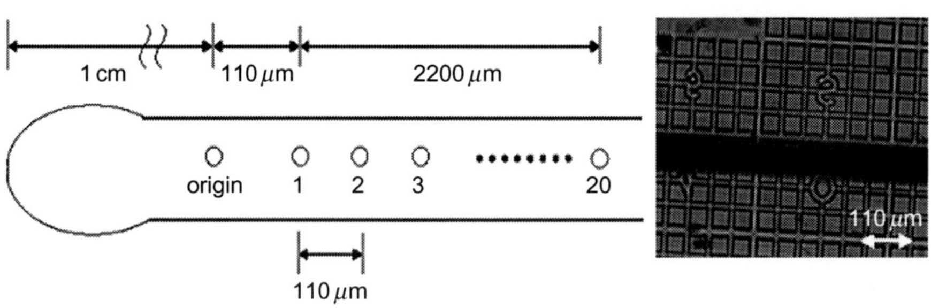

Figure 4 Illustration of the method to measure FTIR spectrum of one hair fiber with FTIR–ATR spectroscopy with microscope.

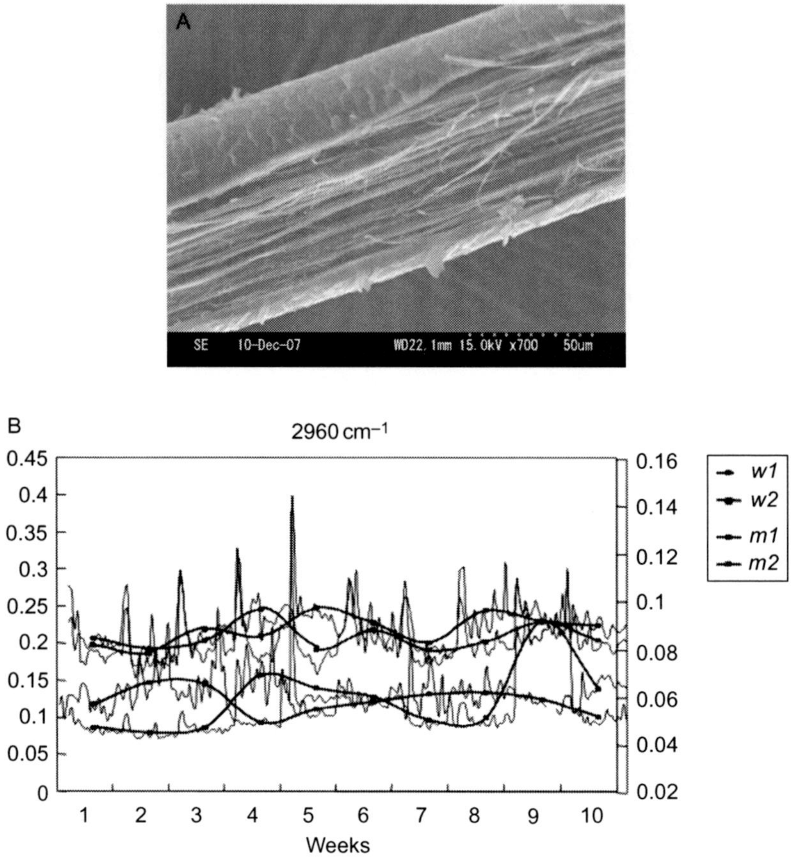

Figure 5 (A) A scanning electron micrograph of half-cut hair. (B) The periodic changes of relative IR intensities at 2960 cm^{-1} in the direction along longitudinal axis of hair fibers for four subjects.

The collection of naturally lost hairs once a week and the FTIR measurement of medulla surface at the fixed position of the hairs may detect time-dependent changes of hair molecules. Figure 6A and B showed time-dependent changes of IR absorption intensities at around 2855 cm^{-1} and changes of dietary total lipids intake which were calculated from the daily-food intake questionnaire with FFQg program (Food Frequency Questionnaire Based on Food Groups in Japan; http://www.kenpakusha.co.jp/), respectively. Then one could search the time gap between the date of lipid intake and the date of appearance of lipids in the hair with higher correlation by changing the time-gap step by step.

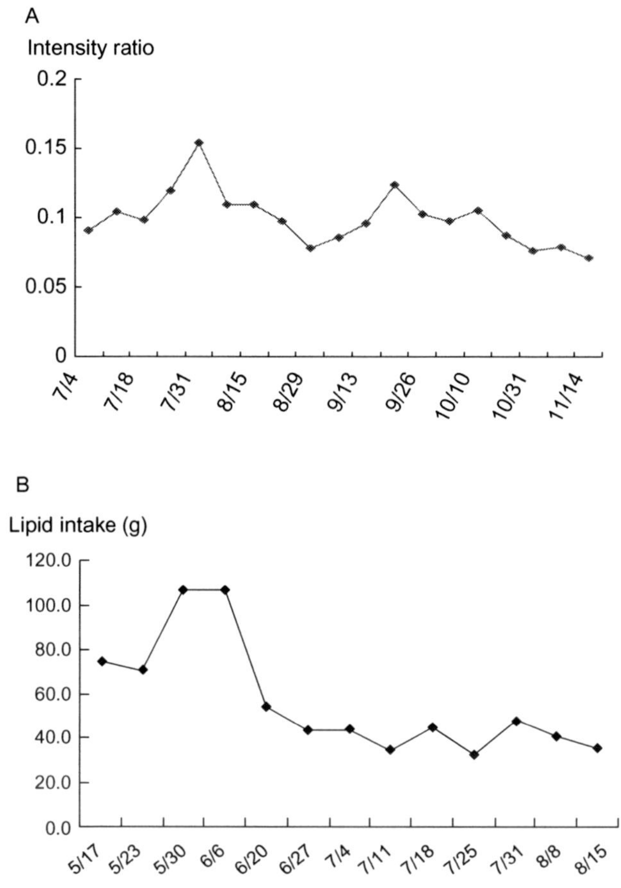

Figure 6 (A) Time-dependent changes of IR absorption intensities at around 2855 cm^{-1}. (B) Time-dependent changes of dietary total lipids intake.

Figure 7 shows the highest correlated case between the hair lipid ratio in FTIR and the amount of lipid intake with significant correlated coefficient ($r = 0.88$), which was found by step-by-step validation, and actually in this case the time gap was 11 weeks.

In Table 1, the summarized data were presented for four male subjects and the corresponding time gaps were nearly 10 or 17 weeks. Interestingly, the negative correlations were also observed for two cases, and this suggested that the accumulation of lipids into hair was observed 10 or 16 weeks later after the intake of lipids on the one hand, and the removal of lipids from hair might be observed on the other hand. In the latter case, the accumulation of lipids may occur in the visceral fat or subcutaneous fat mainly, and the other accumulation in hair or other peripheral tissues may

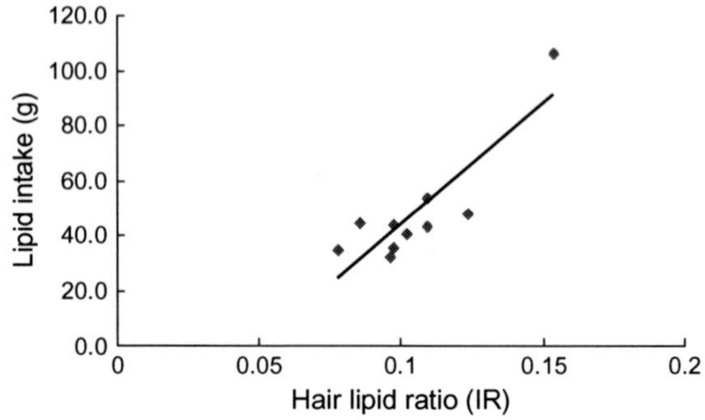

Figure 7 Correlation between the hair lipid ratio in FTIR and the amount of dietary lipid intake.

Table 1 The highest correlation coefficients between the hair lipid ratio in FTIR and the amount of lipid intake for four subjects with the corresponding time gaps

Subject no.	Correlation coefficient (r)	Highly correlated time-gap correspondent (weeks)
1	0.68	8–11
2	0.88	8–11
	0.88	15–18
3	−0.86	15–18
4	−0.73	7–9

be suppressed. Actually, the subject no. 3 and 4 showed the negative correlation and also high body fat values which were independently measured by the electric current sensor (Tanita Co., Japan). Human hair growth cycle was thought to contain normally 6–8 weeks of catagen and telogen phases without hair cell growth and the incorporation of lipids in hair may be regulated only in the anagen (growth) phase. Then, the 10-weeks gap between the date of lipid intake and the calculated date of appearance in the hair may suggest that the measured point in the hair, nearly 10 mm apart from the hair root top, corresponded roughly to 4 weeks, and nearly 6 weeks of catagen and telogen phases may be the resting period and then the hair was naturally released from the scalp skin.

4.4. *In Situ* Measurements of Skin Surface Molecules

The barrier function of the skin is mainly provided by the SC, where CERs, CHOL, and FFAs are present mainly and the lipids in SC form highly ordered crystalline lamellae phase. These lamellae are crucial for a proper skin barrier function. FTIR spectroscopy was used to examine the lipid organization of mixtures prepared from synthetic CERs with CHOL and FFAs. A report [56] showed that conformational ordering and lateral packing of these mixtures indicated great similarities to the lipid organization in SC and lipid mixtures prepared with native CERs. In SC, the number of hydroxy (OH) groups in the head groups of CER subclasses varied. Acyl CERs with a linoleic acid chemically bound to a long acyl chain were also identified. This report revealed that CER head group architecture affected the lateral packing and conformational ordering of the CER:CHOL:FFA mixtures. Furthermore, while the majority of the lipids formed a crystalline packing, the linoleate moiety of the acyl CERs participated in a "pseudo-fluid" phase.

Recently, the other report [55] showed that the FTIR spectroscopic systematic study revealed the hydrated SC lipid barrier model systems composed of an equimolar mixture of a ceramide, free palmitic acid, and cholesterol. The mixing properties with cholesterol and palmitic acid were considered by using perdeuterated palmitic acid and proteated ceramides. Both molecules could be monitored separately using mid-infrared spectroscopy. At physiological relevant temperatures, between 30 and 35 °C, orthorhombic as well as hexagonal chain packing of the ceramide molecules was observed. The formation of chain packing was dependent on lipid hydration, with a decrease in ceramide hydration favoring the formation of orthorhombic hydrocarbon chain packing, as well as temperature. The investigated SC mixtures exhibited a rich polymorphism from crystalline domains with heterogeneous lipid composition to a "fluid" homogeneous phase. The study showed that under physiological conditions (pH 5.5, hydrated, 30–35 °C) of the skin ternary systems composed of an equimolar ratio of ceramides, free palmitic acid and cholesterol might form gel-like domains delimitated by a liquid-crystalline phase boundary. The presented results supported the microstructural arrangement of the SC lipids as suggested by the "domain mosaic model."

The function of skin SC may be basically regulated by not only lipids, but also other biomolecules and those interactions. Especially, hydrophilic biomolecules in SC are playing important roles in maintaining the barrier function.

Lactic acid is a major α-hydroxy acid in natural moisturizer factor (NMF) and found in SC and also in sweat. It was reported recently that the lactic acid and its compound on SC could be monitored noninvasively by FTIR [57], and this report presented that *in situ* and noninvasive measurement of FTIR spectra of human face surface could detect the circa-monthly rhythmic change of the components which included mainly the mixture of magnesium (Mg)–lactate complex and other skin surface soluble components. It was reported that 12% of lactic acid existed in NMF in the skin, and functioned as a part of moisturizer factor in SC. Lactic acid is normally produced in the glycolytic reactions in the tissue cells through reduction of pyruvate with NADH via lactate dehydrogenase (LDH). Thus, the amount of lactic acid may be controlled by changing LDH activity and the amount of NADH and other glucose metabolites in the energy metabolism in human cells. On the other hand, acidic mucopolysaccharides or sulfated glycosaminoglycans (GAGs) play important roles in the extracellular matrix of skin tissues, and these include keratan sulfate and dermatan sulfate. The sulfated GAGs had a sulfate-derived FTIR absorption normally at 1200–1250 cm^{-1} originated from S=O stretching mode [58], and actually a peak at 1243 cm^{-1} could be observed in the component with circa-monthly changes of intensity in the outermost layer of the face skin. On the other hand, the water extract of T1-face skin contained negatively charged fraction (named as Fraction X) in HPLC with a strong anion-exchange column, and this fraction after dried showed FTIR spectrum with broad bands at 1100–1200 cm^{-1}, not clear peak at 1230–1250 cm^{-1}, and this IR absorption pattern was very similar to the IR spectrum of acidic magnesium salt of proteoglycan aggregates [59]. The IR absorption of S=O stretching mode could be largely affected by ionic conditions, especially by binding of a divalent cation with sulfate group. In line with these data, it was suggested that the Fraction X might contain partly magnesium (or calcium)-bound glycosaminoglycan sulfate.

From our present results, Mg–lactate and the aforementioned divalent cation (Mg^{2+})-bound GAGs may play important roles in the circa-monthly rhythmic changes of human skin SC.

In the maintenance and control of formation and exfoliation of skin SC, the activity of protease (cathepsin D) in SC against transglutaminase was reported to play an important role [60]. The turnover of the skin surface cells was reported to have about 4–5 weeks of period [61]. It was reported that in the SC, the endoglycosidase activity of heparanase 1 might be indispensable and represented the first step in the desquamation process [62], and so the degradation of heparan sulfate polymers into shorter chains may be concerned to the exfoliation of SC.

5. Future Perspectives and Conclusion

Infrared spectroscopy, or vibrational spectroscopy, may have a great potential in the application to diagnosis of human diseases. So far, we reviewed mainly the application of mid-infrared spectroscopy to biological cells and tissues, especially the measurements of membranes and lipids and other related biomolecules of human disease-related tissues. Other vibrational spectroscopies, such as near-infrared absorption and Raman scattering spectroscopies, are also important and developing techniques for diagnosis of human diseases, and many applications to biological tissues and cells have been reported.

All these vibrational spectroscopic techniques may have the advantage in measuring changes of biofactors related to diseases as described in this chapter, and many factors including proteins, lipids, glycans, and even other unknown materials may be detected in one shot of experiment, and moreover the interaction among these biofactors in the membranes may also be detected by vibrational spectroscopy. Especially, the latter case is important for the diagnosis of complex human diseases because many factors are actually involved and these factors are more or less connected in the onset of diseases, such as cancer, atherosclerosis, and diabetes mellitus. For example, lowering of LDL–cholesterol in human enhanced the expression of *cyclooxygenase 2* gene in arterial wall and platelet, and changed the inflammation status with changing the production of prostaglandins [63] and thus affected many metabolic status of lipids and other related biomolecules. Changing of one factor would change the state of the other factors and this is the change of metabolic network, and so this network is the target of metabolomics study. This is because the metabolomics study is attracting many medical researches for diagnosis and medical treatment or therapy of diseases.

Vibrational spectroscopies may reveal *nondestructively* and *in situ* the changes of property and quantity of biomolecules of human tissues *in vivo*. Especially for diagnosis of human diseases, the measurement of human skin including lip surface or other easily accessible tissues *in vivo* may be important as skin surface (stratum corneum), may be refreshed nearly every day, and reflects somehow the chemical and biological states of inner tissues which was theoretically and presumably at the stage where cell components were affected several weeks ago depending upon the skin keratinocyte turnover rate. The skin SC may thus contain information about biomolecules including proteins, lipids, and glycans which may be affected several weeks ago in the dermis by the compositions of blood, and especially the diseases such as diabetes and hyper- (or hypo-) lipidemia may change more or less the character of skin biomolecules.

The lipids and membrane structures and functions of human skin SC may be very special, and they play not only the protective role or barrier for human body, but also the regulatory role for reabsorption of exogenous materials such as sweat and sebum components or medicines or other chemicals. The development of more effective drug delivery system through skin tissues (transdermal therapeutic system) may be important in the present and future medical treatments. Infrared spectroscopy, especially FTIR–ATR, could be applied to investigate the mechanism of transdermal transport of chemicals and the enhancer of transport [64,65]. Other vibrational spectroscopic methods like the confocal laser Raman scattering spectroscopic method could also be applied to the skin and transdermal transport research [66].

If the vibrational spectroscopy could be used and handled very easily and precisely to measure the change of biomolecules with sophisticated chemometric methods, it would take the place of the usual clinical tests, especially for screening of diseases, and for this purpose more technical innovations may be needed in the development of infrared spectroscopic machinery specifically for diagnosis of human diseases. The infrared spectroscopic innovations, if succeeded, would result in not only the innovation of clinical screening of human diseases, but also the advancement of realization of the basic metabolic mechanisms of lipids and their interaction with proteins and glycans in human tissues *in vivo*, and after that we would get for the first time the real techniques to monitor physiological molecular dynamics in human.

ACKNOWLEDGMENTS

We thank Dr. S. Sakuyama (Mandom Co., Japan) for help in the skin works. We appreciate the collaboration of many students who carried out many experiments and helped us considerably, namely H. Inukai, Y. Tanaka, T. Yonebayashi, K. Morita, T. Futamura, A. Hayashi, C. Hirabayashi, S. Matsushima, M. Ishibashi, T. Nakata, and M. Yoshida for recent hair and skin works, and Bitoh, Kanai, Kanebayashi, Kayugawa, Miyake, Nakano, and Sugiura for previous hair works. We also thank Mr. H. Tanamachi and Dr. Y. Masukawa (Kao Co., Japan) for their help to write this chapter. We thank Dr. R. Nakayama (Kyoto Women's University) for providing FFQg program. These works were financially supported in part by Grants-in-Aid for Scientific Research from Ministry of Education, Science, Sports, and Culture of Japan to S.Y., and also by Kao Co., Japan. We thank sincerely Dr. Yoshinori Muto (Gifu University) for providing us the chance to write this chapter.

REFERENCES

[1] R.G. Snyder, H.L. Strauss, C.A. Elliger, Carbon-hydrogen stretching modes and the structure of *n*-alkyl chains. 1. Long, disordered chains, J. Phys. Chem. 86(26) (1982) 5145–5150.

[2] R.G. Snyder, G.L. Liang, H.L. Strauss, R. Mendelsohn, IR spectroscopic study of the structure and phase behavior of long-chain diacylphosphatidylcholines in the gel state, Biophys. J. 71(6) (1996) 3186–3198.

[3] B. Cannon, G. Heath, J. Huang, P. Somerharju, J.A. Virtanen, K.H. Cheng, Time-resolved fluorescence and Fourier transform infrared spectroscopic investigations of lateral packing defects and superlattice domains in compositionally uniform cholesterol/phosphatidylcholine bilayers, Biophys. J. 84(6) (2003) 3777–3791.

[4] V.R. Kodati, M. Lafleur, Comparison between orientational and conformational orders in fluid lipid bilayers, Biophys. J. 64(1) (1993) 163–170.

[5] M. Fidorra, L. Duelund, C. Leidy, A.C. Simonsen, L.A. Bagatolli, Absence of fluid-ordered/fluid-disordered phase coexistence in ceramide/POPC mixtures containing cholesterol, Biophys. J. 90(12) (2006) 4437–4451.

[6] S. Yoshida, M. Miyata, H. Yoshida, M. Takeshita, Change of conformational disorder in membrane lipids in the pulmonary artery of monocrotaline-injected rats detected in situ by Fourier transform infrared spectroscopy, Vib. Spectrosc. 3(4) (1992) 271–276.

[7] S. Yoshida, N. Makino, M. Takeshita, Effect of anoxia on carotid of spontaneously hypertensive rat: studies by scanning electron microscopy and Fourier transform infrared spectroscopy, J. Clin. Biochem. Nutr. 8 (1990) 41–50.
[8] S. Yoshida, M. Miyazaki, K. Sakai, M. Takeshita, S. Yuasa, A. Sato, et al. Fourier transform infrared spectroscopic analysis of rat brain microsomal membranes modified by dietary fatty acids: possible correlation with altered learning behavior, Biospectroscopy 3(4) (1997) 281–290.
[9] C.S. Rao, S. Damodaran, Surface pressure dependence of phospholipase A2 activity in lipid monolayers is linked to interfacial water activity, Colloids Surf. B Biointerfaces 34(3) (2004) 197–204.
[10] E.A. Disalvo, F. Lairion, F. Martini, E. Tymczyszyn, M. Frias, H. Almaleck, et al. Structural and functional properties of hydration and confined water in membrane interfaces, Biochim. Biophys. Acta 1778(12) (2008) 2655–2670.
[11] M.C. Luzardo, F. Amalfa, A.M. Nunez, S. Diaz, A.C. Biondi De Lopez, E.A. Disalvo, Effect of trehalose and sucrose on the hydration and dipole potential of lipid bilayers, Biophys. J. 78(5) (2000) 2452–2458.
[12] J.H. Crowe, L.M. Crowe, D. Chapman, Infrared spectroscopic studies on interactions of water and carbohydrates with a biological membrane, Arch. Biochem. Biophys. 232 (1) (1984) 400–407.
[13] J.V. Ricker, N.M. Tsvetkova, W.F. Wolkers, C. Leidy, F. Tablin, M. Longo, et al. Trehalose maintains phase separation in an air-dried binary lipid mixture, Biophys. J. 84(5) (2003) 3045–3051.
[14] D.A. Brown, E. London, Functions of lipid rafts in biological membranes, Annu. Rev. Cell Dev. Biol. 14 (1998) 111–136.
[15] G. Vereb, J. Szollosi, J. Matko, P. Nagy, T. Farkas, L. Vigh, et al. Dynamic, yet structured: the cell membrane three decades after the Singer–Nicolson model, Proc. Natl. Acad. Sci. USA 100(14) (2003) 8053–8058.
[16] Z.D. Schultz, I.W. Levin, Lipid microdomain formation: characterization by infrared spectroscopy and ultrasonic velocimetry, Biophys. J. 94(8) (2008) 3104–3114.
[17] M. Edidin, The state of lipid rafts: from model membranes to cells, Annu. Rev. Biophys. Biomol. Struct. 32 (2003) 257–283.
[18] S.J. Plowman, C. Muncke, R.G. Parton, J.F. Hancock, H-ras, K-ras, and inner plasma membrane raft proteins operate in nanoclusters with differential dependence on the actin cytoskeleton, Proc. Natl. Acad. Sci. USA 102(43) (2005) 15500–15505.
[19] E. London, How principles of domain formation in model membranes may explain ambiguities concerning lipid raft formation in cells, Biochim. Biophys. Acta 1746 (3) (2005) 203–220.
[20] J.R. Silvius, Partitioning of membrane molecules between raft and non-raft domains: insights from model-membrane studies, Biochim. Biophys. Acta 1746(3) (2005) 193–202.
[21] A. Bhushan, M.G. McNamee, Differential scanning calorimetry and Fourier transform infrared analysis of lipid–protein interactions involving the nicotinic acetylcholine receptor, Biochim. Biophys. Acta 1027(1) (1990) 93–101.
[22] D. Dorn-Zachertz, G. Zimmer, Different protein–lipid interaction in human red blood cell membrane of Rh positive and Rh negative blood compared with Rhnull, Z. Naturforsch. C 36(11–12) (1981) 988–996.
[23] F. Barcelo, J. Prades, J.A. Encinar, S.S. Funari, O. Vogler, J.M. Gonzalez-Ros, et al. Interaction of the C-terminal region of the Ggamma protein with model membranes, Biophys. J. 93(7) (2007) 2530–2541.
[24] Y.P. Zhang, R.N. Lewis, R.S. Hodges, R.N. McElhaney, Peptide models of the helical hydrophobic transmembrane segments of membrane proteins: interactions of acetyl-K2-(LA)12-K2-amide with phosphatidylethanolamine bilayer membranes, Biochemistry 40(2) (2001) 474–482.

[25] M. Gasset, J.M. Mancheno, J. Lacadena, A. Martinez del Pozo, M. Onaderra, J.G. Gavilanes, Spectroscopic characterization of the alkylated alpha-sarcin cytotoxin: analysis of the structural requirements for the protein–lipid bilayer hydrophobic interaction, Biochim. Biophys. Acta 1252(1) (1995) 43–52.
[26] I.J. Vereyken, V. Chupin, F.A. Hoekstra, S.C. Smeekens, B. de Kruijff, The effect of fructan on membrane lipid organization and dynamics in the dry state, Biophys. J. 84 (6) (2003) 3759–3766.
[27] N.M. Tsvetkova, I. Horvath, Z. Torok, W.F. Wolkers, Z. Balogi, N. Shigapova, et al. Small heat-shock proteins regulate membrane lipid polymorphism, Proc. Natl. Acad. Sci. USA 99(21) (2002) 13504–13509.
[28] B. Rigas, S. Morgello, I.S. Goldman, P.T. Wong, Human colorectal cancers display abnormal Fourier-transform infrared spectra, Proc. Natl. Acad. Sci. USA 87(20) (1990) 8140–8144.
[29] B. Rigas, P.T. Wong, Human colon adenocarcinoma cell lines display infrared spectroscopic features of malignant colon tissues, Cancer Res. 52(1) (1992) 84–88.
[30] M.A. Cohenford, B. Rigas, Cytologically normal cells from neoplastic cervical samples display extensive structural abnormalities on IR spectroscopy: implications for tumor biology, Proc. Natl. Acad. Sci. USA 95(26) (1998) 15327–15332.
[31] D.E. Maziak, M.T. Do, F.M. Shamji, S.R. Sundaresan, D.G. Perkins, P.T. Wong, Fourier-transform infrared spectroscopic study of characteristic molecular structure in cancer cells of esophagus: an exploratory study, Cancer Detect. Prev. 31(3) (2007) 244–253.
[32] E. Bogomolny, S. Argov, S. Mordechai, M. Huleihel, Monitoring of viral cancer progression using FTIR microscopy: a comparative study of intact cells and tissues, Biochim. Biophys. Acta 1780(9) (2008) 1038–1046.
[33] S. Yoshida, Y. Okazaki, T. Yamashita, H. Ueda, R. Ghadimi, A. Hosono, et al. Analysis of human oral mucosa ex vivo for fatty acid compositions using Fourier-transform infrared spectroscopy, Lipids 43(4) (2008) 361–372.
[34] H.S. Lam, A. Proctor, J. Nyalala, M.D. Morris, W.G. Smith, Quantitative determination of low density lipoprotein oxidation by FTIR and chemometric analysis, Lipids 39(7) (2004) 687–692.
[35] S. Yoshida, H. Yoshida, Noninvasive analyses of polyunsaturated fatty acids in human oral mucosa in vivo by Fourier-transform infrared spectroscopy, Biopolymers 74 (5) (2004) 403–412.
[36] J.R. Beattie, S.E. Bell, C. Borgaard, A. Fearon, B.W. Moss, Prediction of adipose tissue composition using Raman spectroscopy: average properties and individual fatty acids, Lipids 41(3) (2006) 287–294.
[37] A. Ripoche, A.S. Guillard, Determination of fatty acid composition of pork fat by Fourier transform infrared spectroscopy, Meat Sci. 58(3) (2001) 299–304.
[38] N.K. Afseth, V.H. Segtnan, B.J. Marquardt, J.P. Wold, Raman and near-infrared spectroscopy for quantification of fat composition in a complex food model system, Appl. Spectrosc. 59(11) (2005) 1324–1332.
[39] S. Yoshida, Q.Z. Zhang, S. Sakuyama, S. Matsushima, Metabolism of fatty acids and lipid hydroperoxides in human body monitoring with Fourier transform infrared spectroscopy, Lipids Health Dis. 8 (2009) 28.
[40] D.A. Scott, D.E. Renaud, S. Krishnasamy, P. Meric, N. Buduneli, S. Cetinkalp, et al. Diabetes-related molecular signatures in infrared spectra of human saliva, Diabetol. Metab. Syndr. 2 (2010) 48.
[41] M. Sola-Penna, Metabolic regulation by lactate, IUBMB Life 60(9) (2008) 605–608.
[42] M.A. Mackanos, C.H. Contag, Fiber-optic probes enable cancer detection with FTIR spectroscopy, Trends Biotechnol. 28(6) (2010) 317–323.

[43] A. Beljebbar, S. Dukic, N. Amharref, M. Manfait, Screening of biochemical/histological changes associated to C6 glioma tumor development by FTIR/PCA imaging, Analyst 135(5) (2010) 1090–1097.
[44] M.J. Baker, C. Clarke, D. Demoulin, J.M. Nicholson, F.M. Lyng, H.J. Byrne, et al. An investigation of the RWPE prostate derived family of cell lines using FTIR spectroscopy, Analyst 135(5) (2010) 887–894.
[45] V.W. Petit, M. Refregiers, C. Guettier, F. Jamme, K. Sebanayakam, A. Brunelle, et al. Multimodal spectroscopy combining time-of-flight-secondary ion mass spectrometry, synchrotron-FT-IR, and synchrotron-UV microspectroscopies on the same tissue section, Anal. Chem. 82(9) (2010) 3963–3968.
[46] A. Zwielly, S. Mordechai, I. Sinielnikov, A. Salman, E. Bogomolny, S. Argov, Advanced statistical techniques applied to comprehensive FTIR spectra on human colonic tissues, Med. Phys. 37(3) (2010) 1047–1055.
[47] Y. Tuo, P. Huang, Y. Ke, S. Fan, Q. Lu, B. Xin, et al. Attenuated total reflection Fourier transform infrared spectroscopic investigation of the postmortem metabolic process in rat and human kidney cortex, Appl. Spectrosc. 64(3) (2010) 268–274.
[48] L. Zhang, A. Aksan, Fourier transform infrared analysis of the thermal modification of human cornea tissue during conductive keratoplasty, Appl. Spectrosc. 64(1) (2010) 23–29.
[49] M.A. Mackanos, C.H. Contag, FTIR microspectroscopy for improved prostate cancer diagnosis, Trends Biotechnol. 27(12) (2009) 661–663.
[50] U. Zelig, J. Kapelushnik, R. Moreh, S. Mordechai, I. Nathan, Diagnosis of cell death by means of infrared spectroscopy, Biophys. J. 97(7) (2009) 2107–2114.
[51] K. Belbachir, R. Noreen, G. Gouspillou, C. Petibois, Collagen types analysis and differentiation by FTIR spectroscopy, Anal. Bioanal. Chem. 395(3) (2009) 829–837.
[52] Y. Masukawa, H. Narita, G. Imokawa, Characterization of the lipid composition at the proximal root regions of human hair, J. Cosmet. Sci. 56(1) (2005) 1–16.
[53] Y. Masukawa, H. Tsujimura, H. Tanamachi, H. Narita, G. Imokawa, Damage to human hair caused by repeated bleaching combined with daily weathering during daily life activities, Exog. Dermatol. 3(6) (2004) 273–281.
[54] Y. Masukawa, H. Tsujimura, H. Narita, Liquid chromatography-mass spectrometry for comprehensive profiling of ceramide molecules in human hair, J. Lipid Res. 47 (7) (2006) 1559–1571.
[55] P. Garidel, B. Fölting, I. Schaller, A. Kerth, The microstructure of the stratum corneum lipid barrier: mid-infrared spectroscopic studies of hydrated ceramide:palmitic acid:cholesterol model systems, Biophys. Chem. 150(1–3) (2010) 144–156.
[56] M. Janssens, G.S. Gooris, J.A. Bouwstra, Infrared spectroscopy studies of mixtures prepared with synthetic ceramides varying in head group architecture: coexistence of liquid and crystalline phases, Biochim. Biophys. Acta 1788(3) (2009) 732–742.
[57] S. Sakuyama, C. Hirabayashi, J. Hasegawa, S. Yoshida, Analysis of human face skin surface molecules in situ by Fourier-transform infrared spectroscopy, Skin Res. Technol. 16(2) (2010) 151–160.
[58] M.O. Longas, K.O. Breitweiser, Sulfate composition of glycosaminoglycans determined by infrared spectroscopy, Anal. Biochem. 192(1) (1991) 193–196.
[59] S.M. Bychkov, V.N. Bogatov, S.A. Kuz'mina, Analysis of the salts of cartilage proteoglycan aggregates, Biull. Eksp. Biol. Med. 94(11) (1982) 52–55.
[60] F. Egberts, M. Heinrich, J.M. Jensen, S. Winoto-Morbach, S. Pfeiffer, M. Wickel, et al. Cathepsin D is involved in the regulation of transglutaminase 1 and epidermal differentiation, J. Cell Sci. 117(Pt. 11) (2004) 2295–2307.
[61] G. Lindwall, E.A. Hsieh, L.M. Misell, C.M. Chai, S.M. Turner, M.K. Hellerstein, Heavy water labeling of keratin as a non-invasive biomarker of skin turnover in vivo in rodents and humans, J. Invest. Dermatol. 126(4) (2006) 841–848.

[62] D. Bernard, B. Mehul, C. Delattre, L. Simonetti, A. Thomas-Collignon, R. Schmidt, Purification and characterization of the endoglycosidase heparanase 1 from human plantar stratum corneum: a key enzyme in epidermal physiology? J. Invest. Dermatol. 117(5) (2001) 1266–1273.
[63] L.H. Smith, M.S. Petrie, J.D. Morrow, J.A. Oates, D.E. Vaughan, The sterol response element binding protein regulates cyclooxygenase-2 gene expression in endothelial cells, J. Lipid Res. 46(5) (2005) 862–871.
[64] E. Touitou, V.M. Meidan, E. Horwitz, Methods for quantitative determination of drug localized in the skin, J. Control. Release 56(1–3) (1998) 7–21.
[65] Y. Obata, S. Utsumi, H. Watanabe, M. Suda, Y. Tokudome, M. Otsuka, et al. Infrared spectroscopic study of lipid interaction in stratum corneum treated with transdermal absorption enhancers, Int. J. Pharm. 389(1–2) (2010) 18–23.
[66] N. Nakagawa, M. Matsumoto, S. Sakai, In vivo measurement of the water content in the dermis by confocal Raman spectroscopy, Skin Res. Technol. 16(2) (2010) 137–141.

CHAPTER TWO

Mechanical Properties of Bilayer Lipid Membranes and Protein–Lipid Interactions

Tibor Hianik*

Contents

Abstract

Mechanical properties of the lipid bilayers play a crucial role in maintaining the membrane stability, the shape of the cell, in functioning of the protein molecules and are responsible for protein–lipid interactions. Due to structural inhomogeneity, the membrane is characterized by anisotropy of the mechanical properties. Therefore, description of the mechanical properties of the membrane requires the study of membrane deformation in different directions. This chapter reviews current achievements in understanding the membrane mechanics and its role in explanation of the mechanisms of protein–lipid interactions. Several examples of the effect of short peptides and large integral proteins on the mechanical properties of the membrane are presented. We have shown that proteins can affect the ordering of the lipid bilayer on substantially larger distance in comparison with that of annular lipids surrounding of the macromolecule.

* Corresponding author: Tel.: +421 2 60295683; Fax: +421 2 65426774.
E-mail address: tibor.hianik@fmph.uniba.sk

Department of Nuclear Physics and Biophysics, Faculty of Mathematics, Physics and Informatics, Comenius University, Bratislava, Slovakia

Advances in Planar Lipid Bilayers and Liposomes, Volume 13
ISSN 1554-4516, DOI: 10.1016/B978-0-12-387721-5.00002-X

1. Introduction

Protein–lipid interactions play an essential role in the functioning of biomembranes [1,2]. They are responsible for maintaining the membrane integrity and stability and are important for functioning the membrane proteins, such as mechanosensitive ion channels [3] or other integral proteins, for example, bacteriorhodopsin, Na,K-ATPase, Ca-ATPase, etc. The functioning of membrane proteins, which is accompanied by changes in their conformation, could affect the structure and physical properties of the surrounding lipid environment.

For the analysis of the mechanism of protein–lipid interaction, the thermodynamics and mechanics properties of lipid bilayers and proteoliposomes are important. Owing to the possible different geometry of the hydrophobic moiety of proteins and that of lipids, as well as to the action of electrostatic and elastic forces, regions of altered structure may arise around protein molecules [4,5]. The formation of similar regions may represent one of the reasons for the occurrence of long-distance interactions in membranes. Very likely, hydrophobic interactions play the key role in the establishment of links between integral proteins and lipids. The rigid hydrophobic parts of membrane-spanning proteins cause a deformation of the hydrophobic lipid chains due to length matching. This leads to the stretching or compression of the hydrocarbon lipid chains depending on the relation of the hydrophobic part of proteins and the surrounding lipids [4]. Distortion of the membrane by proteins may cause lipid-mediated attractive or repulsive forces between proteins [6]. Due to changes of the ordering of the lipid bilayer, the increase or decrease of phase transition temperature, as well as changes of membrane mechanical properties take place. Due to considerable problems with the isolation and purification of integral proteins and with the determination of their structure, only a few proteins were analyzed so far in respect of their influence on the thermodynamic and mechanical properties of lipid bilayers. Using the mattress model of Mouritsen and Bloom [4] as well as Landau–de Gennes theory of elasticity of liquid crystals, it was possible to explain satisfactorily the changes in the temperature of phase transition, ΔT, of proteoliposomes containing membrane-bound reaction center protein (RC) and antenna protein (LHCP) [7,8]. The ΔT value depends on the mismatch between the hydrophobic part of protein and lipids as well as on the protein concentration according to the following equation [7]:

$$\Delta T = 8\xi^2(2r_0/\xi + 1)\left[2\left(h^{\mathrm{f}} - h_{\mathrm{p}}\right)/\left(h^{\mathrm{f}} - h^{\mathrm{g}}\right) - 1\right]x_{\mathrm{p}}, \qquad (1)$$

where ξ is the characteristic decay length, r_0 is the radius of protein, h^{f} and h^{g} are the length of hydrocarbon chains of phospholipids in a fluid or gel state,

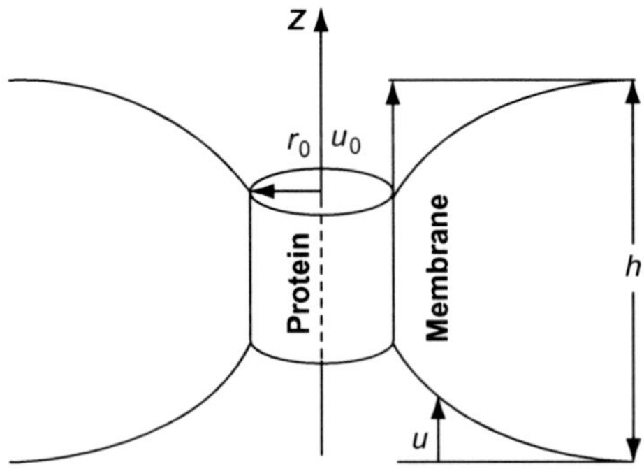

Figure 1 Schematic cross section of an integral protein in a phospholipid membrane: h is the bilayer thickness and r_0 is the radius of the protein.

respectively, h_p is the length of the hydrophobic moiety of protein, and x_p is the molar ratio of protein and phospholipid (number of protein molecules/number of phospholipid molecules). Parameter $\xi(T)$ is not measurable in the experiment and should be determined theoretically. The approach for determination of these parameters was made in work by Huang [9] and Helfrich and Jacobson [10] in which the gramicidin has been used as a model of integral protein. Figure 1 shows a schematic cross section of integral protein in a membrane. Parameter ξ can be determined from the minima of the free energy of the system with the assumption of exponential decay of perturbation:

$$u(r) = u_0 \exp[-(r - r_0)/\xi]. \tag{2}$$

The free deformation energy per unit area of membrane in the cylindrical polar coordinates can be expressed as

$$F = 2\pi \int r\,\mathrm{d}r \left[E_\perp u^2/h + hK_1(u'/r + u'')^2 + \gamma(u')^2\right], \tag{3}$$

where $E_\perp$, K_1, and γ are the elasticity modulus of transversal compression, splay, and surface tension, respectively. To determine the minimum energy conformation, it is necessary to minimize the free energy with respect to the variation in $u(x, y)$ and get the linear differential equation [10]:

$$K_1\left(u'/r^3 - u''/r^2 + 2u'''/r + u^{\mathrm{IV}}\right) - (\gamma/h)(u'/r + u'') + \left(E_\perp/h^2\right)u = 0. \tag{4}$$

Equation (4) can be solved numerically using algorithm described by Pereyra [11]. Thus, theoretical analysis of distorted regions around proteins in the membranes requires knowledge on the elastic parameters of the membrane. The first attempt to characterize the distortion around integral protein bacteriorhodopsin (BR) was made in our work [12] (see also Ref. [5]). We showed that the diameter of distorted region around BR could be up to 40 nm, which is around 10 times more than the diameter of BR (3.5 nm).

The experimental works certainly suggest on existence of distorted regions around proteins. The influence of BR on the structural state of spacious regions of planar bilayer lipid membranes (BLMs) was shown by means of measurement of elasticity modulus $E_{\perp}$ [13]. It was shown that the area of a lipid bilayer with an altered structure per one cluster consisting of three BR molecules surpasses 2800 nm^2. Moreover, as a result of the illumination of BLM modified by BR, a considerable increase of $E_{\perp}$ occurred (more than five times) with a further saturation on a stable level. This condition was preserved for several hours after the illumination was switched off. This shows the possibility of mechanical energy accumulation in membrane. Substantial decrease of volume compressibility of liposomes containing BR was demonstrated by the ultrasound velocimetry method [14]. In particularly, it was shown that one BR trimer is able to change the structural state of the lipid bilayer of large unilamellar liposomes composed of dipalmitoylphosphatidylcholine (DPPC) and within a diameter of approximately 100 nm.

The above results have been obtained with so-called macroscopic methods which "see" large membrane area [5]. The microscopic methods such as EPR, NMR, or fluorescence spectroscopy are also informative in the study of protein–lipid interactions. However, these methods usually "see" the changes in lipid bilayers in a close proximity of the protein. The EPR spectra reveal a reduction in mobility of the spin-labeled lipid chains on the binding of peripheral proteins to negatively charged lipid bilayers. Integral proteins induce a more direct motional restriction of the spin-labeled lipid chains, allowing the stoichiometry and specificity of the interaction, and the lipid exchange rate at the protein interface, to be determined by EPR spectra. In this way, a population of very slowly exchanging cardiolipin associated with the mitochondrial ADP–ATP carrier has been identified (see Ref. [15] for review). However, the exchange between the lipids adjacent to the proteins and those in a bulk phase is in general fast [16]. Fluorescence spectroscopy is also effective for the study of protein–lipid interactions. In particular Rehorek *et al.* [17] showed that as a result of conformational changes of the BR, the ordering of the lipid bilayer increases and a transmission of conformational energy occurs over the distance more than 4.5 nm.

Biophysical studies of the mechanisms of protein–lipid interactions suggest on the existence of annular lipids surrounding the integral membrane proteins [2]. These lipids are, however, not simply adjacent to the protein, but as we mentioned earlier, exhibit in fast exchange with bulk lipids [16]. The specificity of these interactions is, however, under discussion. Most probably, the exact lipid composition is not so essential for the protein–lipid interactions [2]. For example, changes of the fatty acid chain composition caused by diet have no injurious effect on cell function. However, dietary-induced changes in lipid composition are limited: some features of the fatty acyl chain composition are maintained constant, like chain length in the range between, typically C16 and C20, with about half the chains being saturated and half unsaturated. This means that overall features of the lipid composition, such as length and saturation of fatty acid, are likely to be important for the membrane properties [2]. The length and saturation of fatty acid are responsible for the creation of certain thickness of the membrane and its physical state, which are important factors that determine the protein–lipid interactions [1,2]. In certain case, the structure of the polar part of the phospholipids also plays a role in protein–lipid interactions. There is, however, evidence that small number of special lipids is important for the function of the protein. For example, for Ca^{2+}-ATPase, it is phosphatidylinositol-4-phosphate. The binding of this lipid resulted in increasing the activity of the calcium pump by twice [2]. Specific activity to cardiolipin has been observed for cytochrome *c* oxidase [16]. At the same time, the lipid packing also affects the protein–lipid interactions [18]. Due to inhomogeneities of the protein surface, the surrounding lipids tend to adapt their conformation in order to gain optimal contact with the hydrophobic surface of the protein. It has been established that, for example, in the case of Ca^{2+}-ATPase, approximately 30 lipid molecules surround the protein in an annulus-like ring [19]. The lipid environment affects the functioning of ATPase. Starling *et al.* [20] found differences in the rate of dephosphorylation of the Ca-ATPase depending on the thickness of the bilayer, with the slowest rate for di(C24:1)PC bilayers. The role of annular lipids in the protein–lipid interactions has been reviewed recently in paper by Ernst *et al.* [21].

Beyond existing observations, the question arises whether only annular phospholipids are involved in protein–lipid interactions, or whether a larger environment of the lipid bilayer is affected by the presence of an integral protein. In addition, it is necessary to analyze how the properties of the lipid bilayers are altered in close proximity of the proteins compared to the unperturbed lipid bilayer. Despite extensive experimental and theoretical works in the study of the protein–lipid interactions, the understanding of the mechanisms of this process is not yet complete. For example, the aforementioned mattress model of protein–lipid interaction proposed by Mouritsen and Bloom [4] is based on the mismatch between the hydrophobic part of integral protein and the lipids. The mismatch resulted in changes of phase transition temperature of the phospholipids. Typical situation is presented schematically in Fig. 2. If the hydrophobic thickness of the

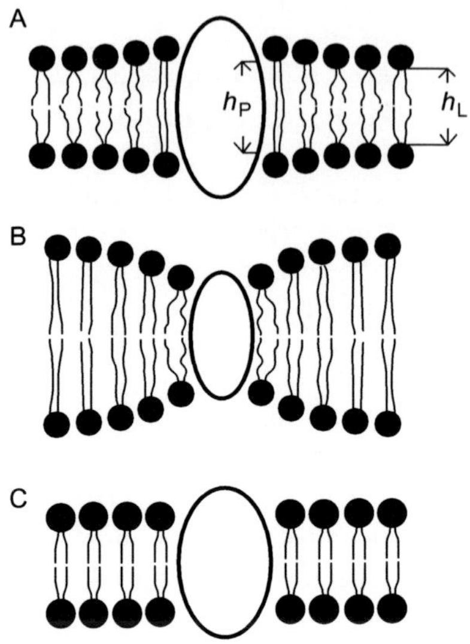

Figure 2 The possible effect of proteins on the structural state of lipid bilayer depending on the mismatch between hydrophobic length of protein, h_P, and lipids, h_L. (A) $h_P > h_L$, (B) $h_P < h_L$, and (C) $h_P = h_L$.

protein h_P is larger than that of the lipids (h_L), the protein induces ordering region of the membrane in its proximity. As a result, the phase transition temperature increases (Fig. 2A). In contrast, when $h_P < h_L$, the disordered region around protein appears and the phase transition of lipids decreases (Fig. 2B). When $h_P = h_L$ (Fig. 2C), the lipid bilayer is not disturbed and thus, no changes in thermodynamic parameters are expected. However, even in the later case, we observed substantial affect of BR on the mechanical properties of the lipid bilayers composed of DPPC with the hydrophobic length similar to that of BR [14]. The molecular dynamic (MD) simulation studies revealed that except of distortion of the lipid bilayer geometry around the protein also certain tilt of the peptide should take place in the membrane with hydrophobic thickness lower in comparison with those of the protein [22,23]. There is also certain discrepancy between microscopic methods, like ESR and NMR, which suggest on effect of protein on to closely adjacent phospholipids and membrane mechanics that evidence on appearance rather large distortion membrane regions that surpass the protein diameter.

In this chapter, we focus on application of membrane mechanics to study the protein–lipid interactions. Membrane mechanics is, however, rather complex. Due to inhomogeneity and anisotropy of membrane, the mechanical properties are also anisotropic and full characterization requires description of deformation of the membrane in different directions.

Therefore in first part, we present short introduction into the anisotropy of the membrane mechanical properties. In second part, we discuss the application of special method—the ultrasound velocimetry—that allows to study volume compressibility of the membrane modified by short peptides and Na,K-ATPase.

2. Mechanical Properties of BLMs and Their Anisotropy

Viscoelastic properties have a significant role in allowing biomembranes to perform different functions. Together with cytoskeleton, viscoelasticity determines the cell shape and transduction of mechanical deformation from mechanoreceptors to sensitive centers. In addition, during conformational changes of the proteins, also the physical properties of the membrane could change. These changes could be described by means of macroscopic approach using the theory of elasticity of solid bodies and liquid crystals [5]. The structure of lipid bilayer is considerably simpler than that of biomembranes. However, even the structure of BLM has clearly expressed anisotropy. This leads to strong anisotropy of membrane viscoelastic properties and requires the description of bilayer properties by several elasticity moduli. The difficulties of describing the membrane elasticity are still not exhausted by this phenomenon. The behavior of deformable solid bodies is described by the theory of elasticity. The principal differences of biomembranes from the classical objects of this theory are as follows. The membrane thickness is very small, approximately 20–200 atomic dimensions. Therefore, the influence of microheterogeneity of each atomic layer on membrane properties can be substantial. In this case, the macroscopic parameters of the membrane, which are the result of the average of their properties over the environment, can considerably differ from the corresponding parameters of the membrane on the level of some distinguished layers. On the other hand, 20–50 atomic layers are too large a value to enable their description by equations of the theory of elasticity for each layer. Therefore, there exist a number of models of biomembranes as elastic bodies, which average the properties of biomembranes by certain numbers of such layers [24–28]. However, for subsequent analysis, it is necessary to introduce a macroscopic description of membrane as an elastic body and generalize it to account for its viscous properties. This analysis has been performed in monograph by Hianik and Passechnik [5]. It has been shown that understanding of the membrane as a viscoelastic body requires analysis of membrane deformation in different directions (Fig. 3) (1) volume compressibility, (2) area compressibility, (3) unilateral extension along membrane plane, and (4) transversal compression. The mechanical

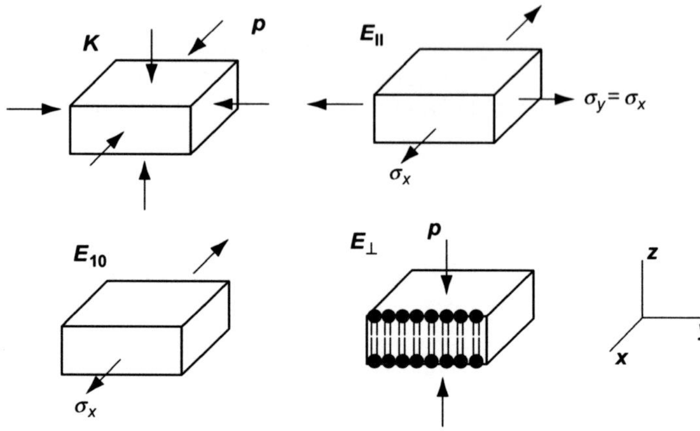

Figure 3 Schematic representation of membrane deformation. Arrows indicate the application of mechanical stress. For explanation, see the text.

parameters that characterize membrane deformability listed earlier are volume compressibility modulus K and Young's moduli of elasticity $E_{\|}$, E_{10}, and $E_{\perp}$, respectively. These parameters are defined as follows:

$$\begin{aligned} &K = -p/(\Delta V/V), \quad E_{\|} = \sigma_x/(\Delta S/S) = 2\sigma_x/(\Delta C/C), \\ &E_{10} = \sigma_x/U_{xx}, \quad E_{\perp} = -p/U_{zz} = 2p/(\Delta C/C), \end{aligned} \tag{5}$$

where σ_x is the mechanical stress along the membrane plane; p is the pressure compressing the membrane; U_{zz} and U_{xx} are the relative membrane deformation in transversal direction and along the membrane plane; and $\Delta V/V$, $\Delta S/S$, and $\Delta C/C$ are the relative changes of the volume, area, and electrical capacitance, respectively. The Young's modulus of area compressibility is related to the bending modulus of elasticity, κ, by simple equation: $\kappa = hE_{\|}$, where h is the membrane thickness [29]. Owing to small dimensions of the membrane, special methods were developed for measurement of the elasticity moduli. Below, we briefly describe the basic methods of measurement of the elasticity moduli of lipid bilayer and show their typical properties. Mechanical properties of lipid bilayers have been described in detail elsewhere [5,30].

Transversal elasticity modulus, $E_{\perp}$. Transversal deformation, that is, parameter $U_{zz} = \Delta h/h$ (h is the membrane thickness) cannot be measured directly due to small thickness of the membrane and extremely small changes of the thickness upon deformation. Therefore, the transversal deformation is determined mostly from the measurement of changes in the electrical capacitance of the membrane. In the case of isovoluminous deformation, that is, when the volume compressibility, K, is much higher than $E_{\perp}$ (this has been certainly shown experimentally [5]), $\Delta h/h = -\Delta C/2C$ (i.e., decrease of thickness resulted in increase of membrane capacitance). In transversal direction, the membrane cannot be deformed by mechanical pressure. However, because the membrane behaves electrically as a capacitor, when voltage is applied to the BLM, it will compress the membrane with an electrostriction pressure $p = C_s U^2/2h$ (C_s is the specific capacitance of the membrane $C_s = C/S$ and U is the applied voltage). Therefore, $E_{\perp} = -p/(\Delta h/h) = 2p/(\Delta C/C)$. For the measurement of changes of capacitance, special method is also required. This is connected with inhomogeneity of the membrane and with the presence of thick Plateau-Gibbs border. For example, if the dc voltage will be applied to BLM and the capacitance will be measured, for example, by capacitance meter, then the measured changes $\Delta C/C$ will be not only due to the changes of the thickness, but also other factors, such as rebuilding of new bilayer parts from the Plateau-Gibbs border will contribute to this value. As a result, the determined elasticity modulus will be underestimated in comparison with their real value. This particularly explains the underestimated values of transversal elasticity modulus in earlier works (see Ref. [5] for review). Therefore, special electrostriction method based on measurement of the amplitude of higher current harmonics has been developed [31]. This method as well as its application to various BLM systems is described in detail in Ref. [5]. Briefly, if an alternating voltage of amplitude U is applied to the BLM through electrodes (e.g., Ag/AgCl electrodes), due to the nonlinear dependence of capacitance on the voltage ($C = C_0(1 + \alpha U^2)$, where C_0 is the capacitance at $U = 0$ and α is the electrostriction coefficient), the higher current harmonics with frequencies $2f$, $3f$, etc., amplitude I_2, I_3, etc., respectively, will be generated in addition to the basic first current harmonic (frequency f) of an amplitude I_1. The measurements of these amplitudes allowed us to determine various parameters, particularly the absolute value of elasticity modulus:

$$E_{\perp} = 3C_s U_0^2 I_1/(4dI_3), \tag{6}$$

where C_s is the specific capacitance of the membrane. If in addition to the amplitude also the phase shift, φ, between first and third current harmonic is measured, then coefficient of dynamic viscosity, η, can be determined:

$$\eta = E_{\perp} \sin\varphi/(2\pi f) \tag{7}$$

(see Ref. [5] for detailed description of the method and experimental setup). Using this method, it has been shown that elasticity modulus and dynamic viscosity depend on the frequency of deformation and increase with the frequency. The value of $E_{\perp}$ also depends on the content of hydrocarbon solvent in a BLM and increases with decreasing the concentration of the solvent, reaching the value of 3.6×10^8 Pa for solvent-free membranes composed of glycerol monooleate at frequency of deformation 1.3 kHz. This modulus is also extremely sensitive to the lipid composition and content of cholesterol or other sterols. For example, $E_\perp$ value increases with increasing the length of hydrocarbon chains of phospholipids that evidence on higher order of hydrophobic part of the membrane due to more extensive hydrophobic interaction between the phospholipid chains [5]. On the other hand, $E_\perp$ decreases with increasing the degree of unsaturation of fatty acids that evidence on decrease the membrane ordering [32]. This elasticity modulus changes considerably upon interaction with BLM of low molecular compounds, for example, local anesthetic or macromolecules, for example, integral or peripheral proteins [5]. The method of measurement $E_\perp$ has been applied also to the supported bilayer lipid membranes (sBLMs) and gave possibility to study the affinity interactions [33] or the interaction with BLM of nucleic acids and their complexes with cationic surfactants [34].

Recently, the excitation FTIR spectroscopy has been applied to study electrostriction of supported lipid multilayers composed of dimyristoylphosphatidylcholine (DMPC) [35]. A periodic rectangular electric potential induced periodic variation of the tilt angle of the hydrocarbon chains by $0.09 \pm 0.015°$. This corresponds to a variation of the bilayer thickness of $\Delta h = 5.4 \times 10^{-3}$ nm. The calculated Young's elastic modulus $E_\perp = 1.8 \times 10^6$ Pa was in good agreement with data obtained by electrostriction method [33]. The electrostriction-induced changes of the tilt angle of acyl chains of DMPC were studied in detail also by Lipkowski and coworkers [36].

The area expansion modulus, $E_{||}$. The area expansion modulus and consequently also the bending elasticity modulus, κ, can be determined by the method of micropipette pressurization of giant bilayer vesicles [37], or by determination of the changes of electrical capacitance during periodical deformation of spherical BLM [25]. The value of $E_{||}$ can be measured only in a limited range of frequencies (5–10 Hz). The typical values of $E_{||}$ of BLM with the hydrocarbon solvent were in the range 10^7–10^8 Pa, which is more than 10 times higher than the values of $E_\perp$ for similar BLM composition at lowest frequency of deformation (20 Hz). The area expansion modulus is less sensitive to the lipid composition and does not significantly depend on the length of hydrocarbon chains of phospholipids and degree of their unsaturation [38].

The elasticity modulus, E_{10}. The elasticity modulus has been measured upon longitudinal distension of cylindrical BLM formed between two circles, with one of them oscillating and the other one being attached to an ergometer. Values $E_{10} \geq 10^6$ Pa have been obtained for membranes of various compositions. They were independent of frequency over an interval of 30–200 Hz, that is, they are determined by bilayer elasticity rather than viscosity [5].

Modulus of volume compressibility, K. Modulus of volume compressibility has been measured by determination of sound velocity in suspension of small unilamellar liposomes. Values of $K = (1.70 \pm 0.17) \times 10^9$ Pa have been determined by this method for liposomes composed of egg phosphatidylcholine (PC). The values of a similar order have been obtained also on large unilamellar liposomes composed of polyunsaturated fatty acids [32]. Using the measurement, the elasticity modulus K, the mechanic and thermodynamic properties of liposomes of various composition, and that contained cholesterol [39] or modified by proteins [40,41] can be studied.

The experiments on determination of various elasticity moduli revealed that these values can be estimated only in a limited range of frequencies: $E_{\|} = 5$–10 Hz, $E_{10} = 2$–300 Hz, $E_{\perp} = 20$ Hz–15 kHz, and $K = 7$ MHz [20]. However, these values can be approximated to the frequency range 10–200 Hz [5]. It has been shown that following inequalities hold for these elasticity moduli: E_{10}, $E_{\perp} \ll E_{\|} \ll K$?. Thus, the BLM represents anisotropic viscoelastic body. The corresponding model of BLM deformation should fulfill the above inequalities.

It has been shown [5] that the mechanical properties of BLM cannot be described by isotropic mechanical models by Wobschall [24] or by Evans and Skalak [42]. Recently discussed brush model of the membrane mechanics composed of two isotropic layers well describes the behavior of area expansion and bending elasticity moduli [38]. However, the model does not provide information about distribution of chain across bilayer and thus, does not consider anisotropy of mechanical properties in transversal direction. The three layer model of deformation has been assumed for description of BLM anisotropy by Passechnik [28]. The two outer layers (thickness h_1) have a modulus of elasticity $E^{(1)}$ and the inner layer (thickness h_2) has a modulus of elasticity $E^{(2)} \ll E^{(1)}$, like a sandwich. Mechanical stress in membrane plane (measurement of $E_{\|}$) deforming the layers with a large elastic modulus ($E^{(1)}$) and the stress perpendicular to the membrane plane (measurement of $E_{\perp}$) deforming the "soft" layer modulus ($E^{(2)}$). Therefore, one can expect that $E_{\perp} \ll E_{\|}$. Deformation of the "sandwich" depends on the degree of adhering of layers. However, full description of deformation requires also consideration of microinhomogeneities in the membrane (see Ref. [5] for more details).

The study of the mechanisms of protein–lipid interactions requires selection of convenient model system. In this respect, the free standing BLMs have been widely used. Most focus was on the study of the properties of the ionic channels such as gramicidin [43] and ionic carrier valinomycin [44]. Later, the BLMs modified by membrane fragments containing insulin receptors or BR [5] were used to study the effect of conformational changes of proteins on the mechanical properties of lipid bilayer. However, BLMs are rather fragile; therefore in further experiments, the sBLMs were used to study the effect of short peptides on the electrostriction of the lipid bilayer

[45]. Recent achievements in fabrication of filter-supported membranes and membranes formed on nanopores (see Ref. [46] for review) are of large promise for future studies of the physical properties of the lipid bilayers at presence of proteins. However, it is undoubtedly that technologically most easier and convenient objects in membrane science are the uni- or multilamellar liposomes. The liposomes could be modified by proteins and so-called proteoliposomes can be prepared [47]. The properties of proteoliposomes can be studied by large number of powerful physical methods, including light scattering, neutron and X-ray diffraction, various optical methods, NMR, ESR, and others. It is important that liposomes are convenient also for the determination of area expansion and bending elastic modulus [42] as well as for the study of the volume compressibility of lipid bilayers [5].

In this work, we focus on determination of volume compressibility of the lipid bilayers with incorporated short peptides or integral proteins. Therefore, we briefly describe the method of measurement volume compressibility using ultrasound velocimetry and densitometry methods.

2.1. Determination of Volume Compressibility by Ultrasound Velocimetry and Densitometry Methods

The measurement of ultrasound velocity allows us to evaluate the elastic properties of aqueous media and suspensions such as liposomes [32,39], lipoproteins [48], or cell surface proteins [49] based on a simple relationship:

$$\beta_S = \frac{1}{\rho u^2}, \tag{8}$$

where β_S, ρ, and u are the adiabatic compressibility, the density, and the sound velocity of the suspension, respectively. Thus, by measuring the changes of sound velocity and density, one can determine the changes of adiabatic compressibility [50].

Ultrasound velocity was measured using a fixed-path differential velocimeter consisting of two almost identical acoustic cavity resonators [50,51] operated at frequencies around 7.2 MHz. The resonance frequencies of the cells were measured using a computer-controlled network analyzer (USAT, USA). The sample volume was 0.7 ml. The resonator cells were equipped with magnetic stirrers to ensure homogeneously dispersed samples during the measurements. One resonator contained the vesicle solution whereas the other one was filled with the same buffer solution without vesicles as reference. When starting a series of measurements, first the resonance frequencies of both resonators were compared by measuring both cells with identical reference liquid. As the energy density of the sonic signal was small throughout (the pressure amplitude in the ultrasonic wave was less than 10^3 Pa), any effects of the sound wave on the structural properties of the vesicles were avoided. In general, ultrasonic velocimetry allows the determination of the sound velocity $[u]$ or rather its concentration increments [51] as defined by the equation:

$$[u] = \frac{u - u_0}{u_0 c}, \tag{9}$$

where c is the solute concentration in mg/ml and the subscript "0" refers to the solvent (buffer). The value $[u]$ can be directly determined from the changes of resonance frequencies f and f_0 of both resonators (f is the resonance frequency of the sample, and f_0 that of the reference-buffer):

$$[u] = \frac{u - u_0}{u_0 c} = \frac{f - f_0}{f_0 c}(1 + \gamma) \tag{10}$$

(the coefficient fulfills the condition $\gamma \ll 1$ [52] and can be neglected in the calculations).

A high precision densitometric system (DMA 60 with two DMA 602 M sample chambers, Anton Paar KG, Graz, Austria) operating according to the vibrating tube principle [53] was used to determine the density ρ of the vesicle solution. Apparent specific partial volumes φ_V have been calculated from the density data using the relation:

$$\varphi_V = \left[1 - \frac{\rho - \rho_0}{c}\right]\frac{1}{\rho_0} = \frac{1}{\rho_0} - [\rho], \tag{11}$$

where the subscript 0 refers again to the reference solvent and $[\rho] = (\rho - \rho_0)/(\rho_0 c)$ denotes the concentration increment of density. The temperature of the cells was controlled to within ± 0.02 °C with a Lauda RK 8 CS ultra-thermostat (Lauda, Germany).

The determination of the specific volume in addition to the sound velocity concentration increment allowed the estimation of the normalized apparent specific compressibility, φ_K/β_0, of the vesicles, which is based on the following equation:

$$\frac{\varphi_K}{\beta_0} = -2[u] - \frac{1}{\rho_0} + 2\varphi_V, \tag{12}$$

where β_0 is the coefficient of the adiabatic compressibility and ρ_0 is the buffer density [51]. The value of φ_K/β_0 indicates the volume compressibility of the vesicles relative to the buffer.

3. The Mechanical Properties of Lipid Bilayers with Incorporated Proteins

In this section, we focus on effect of short peptides and Na,K-ATPase on the volume compressibility of liposomes. The short peptides affect the mechanical properties of the membrane and could model the integral membrane proteins. The distorted regions of the membrane around short peptides caused by the mismatch between lipids and peptides particularly agreed with MD simulations. However, MD method revealed additional alternative of the peptide–membrane matching—the tilt of the peptide in a membrane.

3.1. Effect of Short Peptides on the Mechanical Properties of the Lipid Bilayers

The study of the mechanisms of protein–lipid interactions in native biological systems is rather difficult due to large heterogeneity of both the lipids and proteins. To overcome the problem of the complicated structure of integral proteins and the problems with their isolation and purification, a number of workers have designed and synthesized peptide models (10–40 amino acids) of specific regions of natural membrane proteins and have studied their interactions with model lipid membranes of defined composition (see Refs. [54,55]). In particular, the study of the mechanisms of the interactions of peptides with BLMs has also very important practical significance for understanding of the mechanism of interaction of, for example, neuropeptides [56] or antimicrobial peptides (AMPs) [57] with membranes.

The synthetic peptide acetyl-K_2-G-L_{24}-K_2-A-amide (P_{24}) and its structural analogs, for example, acetyl-K_2-L_{24}-K_2-amide (L_{24}), have been successfully utilized as a model of the hydrophobic transmembrane α-helical segments of integral proteins [55,58]. These peptides contain a long sequence of hydrophobic leucine residues capped at both the N- and C-termini with two positively charged lysine residues. The central polyleucine region of these peptides was designed to form a maximally stable α-helix which will partition strongly into the hydrophobic environment of the lipid bilayer core, while the dilysine caps were designed to anchor the ends of these peptides to the polar surface of the BLM and to inhibit the lateral aggregation of these peptides. In fact, circular dichroism (CD) [58] and FTIR [59] spectroscopic studies of P_{24} have shown that it adopts a very stable α-helical conformation both in solution and in lipid bilayers, and X-ray diffraction [60], fluorescence quenching [61], FTIR [59], and deuterium nuclear magnetic resonance (^{2}H-NMR) [62] spectroscopic studies have confirmed that P_{24} and its analogs assume a transbilayer orientation with the N- and C-termini exposed to the aqueous environment and the hydrophobic polyleucine core embedded in hydrocarbon core of the BLM when reconstituted with various PCs [63]. ^{2}H-NMR [64] and EPR [65] spectroscopic studies have shown that the rotational diffusion of P_{24} about its long axis perpendicular to the membrane plane is rapid in the liquid-crystalline state of the bilayer.

Detailed DSC, FTIR, NMR, and EPR studies of interaction of P_{24} or L_{24} with BLM [66] have revealed that the results obtained from different physical techniques generally agree well with one another. However, certain discrepancies have been found in comparison of the results obtained by spectroscopic techniques, that is, FTIR and ^{2}H-NMR. While the ^{2}H-NMR technique indicated that incorporation of P_{24} peptide into the DPPC bilayers resulted in a decrease of the ordering of the membrane in gel state and increase in the liquid-crystalline state, FTIR experiments suggest that peptide induced a decrease of the ordering of the lipid bilayer in both structural states of the membrane [66]. This discrepancy has been explained by different peculiarities of these two methods. While the order parameters in ^{2}H-NMR spectroscopy are primarily sensitive to *trans/gauche* isomerization, the molecular interpretation of the changes in membrane ordering based on changes in frequency of the methylene stretching modes in IR spectroscopy is likely attributed to the sensitivity of the band position phenomena other than *trans/gauche* isomerization, such as the interchain coupling and the contribution of peptide in the methylene and methyl stretching region. Interchain coupling is significant enough even in fluid bilayers [66]. Therefore, using exclusively FTIR, it is difficult to decide what process is dominant in fluid state-interchain coupling or *trans/gauche* isomerization.

In contrast with spectroscopic methods that provide information about microscopic changes of the lipid bilayer in close proximity of the protein, macroscopic methods, such as membrane compressibility measurements, are sensitive to changes of large membrane regions. The sensitivity and utility of measurements of volume compressibility have been proved in several studies of the interaction of integral proteins with lipid bilayers, for example, BR [5] or peptides like $ACTH_{24}$ [56] or gramicidin S (GS) [57].

Recently, the study of the effect of another short peptide-magainin (MAG) on the mechanical properties of giant unilamellar vesicles (GUVs) composed of palmitoylphosphatidylcholine (POPC)-has been reported [67]. MAG is AMP composed of 23 amino acids. It is produced by African frog *Xenopus laevis*. MAG has antibacterial, antifungal, and anticancer activity and causes leakage of lipid vesicles which is evidence of its perturbing action on lipid bilayer [68,69]. At neutral pH, the peptide carries three net positive charges and contains several hydrophobic residues. At presence of rather low peptide content in the membrane ($<1\%$ of the surface area coverage), the binding rigidity is reduced [67]. It has been suggested that the MAG induces highly mobile regions in the membrane monolayer with high curvature on the membrane surface (see also Ref. [69] for recent review of the effect of AMPs on the membrane properties). The MD simulations of MAG in POPC bilayers indicate that peptide interacts with polar part of the membrane by forming hydrogen bonds with oxygens of phosphate or glycerol. Peptide caused only slight disordering of the bilayer because the hydrocarbon chains that were far from the peptide have not been affected. In addition, MAG caused local thinning of the membrane [70,71]. Simulation studies based on insertion of antibacterial peptides into anionic and zwitterionic lipids suggest that these lipids could modulate the helical structure of the peptide even in larger extend than the perturbation of the membrane induced by peptides [69].

The association of the proteins with the membranes can be enhanced by means of hydrophobic anchor. This is important for association with special membrane structures such as lipid domains, rafts, and caveolae [72]. As an anchor, the hydrophobic chains such as acyl (myristoyl, palmitoyl) or prenyl (linear polyisoprene) are used. Based on fluctuation analysis of GUV, Rowat *et al.* [73] showed that short peptides modified by farnesyl anchor affect the bending rigidity depending on the peptide charge. The charged peptides modestly decreased the main gel–fluid phase transition and markedly increased the bending rigidity of GUV from DMPC membranes, while uncharged species dramatically decreased the main phase transition and modestly decreased the bending rigidity.

Among AMP, GS has also been extensively studied. Cyclic GS is decapeptide first isolated from *Bacillus brevis* [74]. This peptide exhibits appreciable antibiotic activity against a broad spectrum of gram-negative and gram-positive bacteria as well as against several pathogenic fungi [75,76]. In aqueous solution, GS forms an amphiphilic, two-stranded, antiparallel β-sheet structure [75]. Considerably evidence exists that the primary target of GS is the lipid bilayer of cell surface membranes and that this peptide kills cells by destroying the structural integrity of the lipid bilayer [77]. GS partitions strongly into liquid-crystalline lipid bilayers in both model and biological membranes and seems to be located primarily in the glycerol backbone region below the polar head group and above the hydrocarbon chains [78]. In addition, GS binds more strongly to negatively charged lipids [78,79] than to zwitterionic or uncharged phospho- and glycolipids [80,81]. The effect of GS on volume compressibility of large unilamellar vesicles (LUVs) and multilamellar vesicles (MLVs) has been studied in our previous work using ultrasound velocimetry method [40]. The vesicles were prepared from DMPC with different molar ratios of GS and studied the changes of ultrasound velocimetry and specific volume as a function of temperature. These methods, if used simultaneously to study peculiarities of the phase transition of lipid bilayer, allow one to determine the phase transition temperature and the degree of phase transition cooperativity. In addition to the thermodynamic properties of the lipid phase transition obtained by differential scanning calorimetry (DSC), the ultrasound velocimetry and densitometry allow one to study the mechanical properties of lipid bilayer membranes. The basic values of the parameters usually determined in sound velocity and density experiments are shown in Fig. 4 for suspensions of LUVs

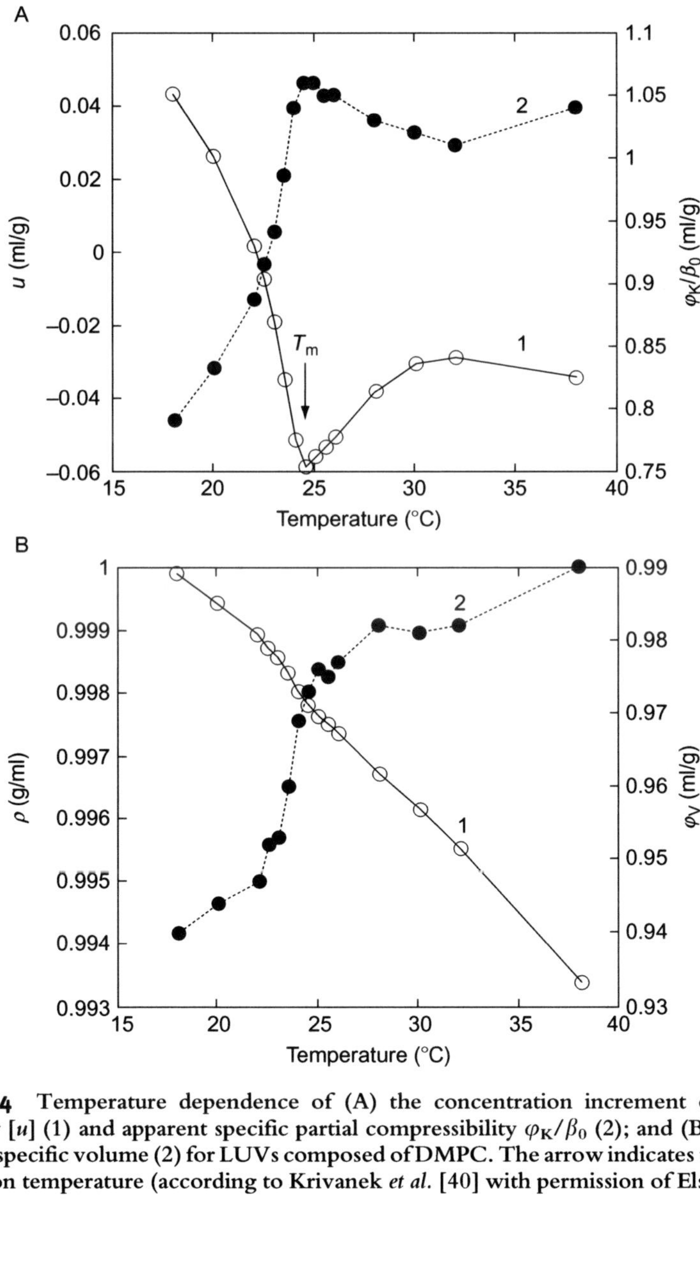

Figure 4 Temperature dependence of (A) the concentration increment of sound velocity [*u*] (1) and apparent specific partial compressibility φ_K/β_0 (2); and (B) density (1) and specific volume (2) for LUVs composed of DMPC. The arrow indicates the phase transition temperature (according to Krivanek *et al.* [40] with permission of Elsevier).

composed of DMPC as a function of temperature and in the range of the phase transition temperature ($T_{\mathrm{m}} \sim 24\ ^{\circ}\mathrm{C}$).

To better understand the obtained data, we shortly explain the physical meaning of ultrasound velocimetry and densitometry parameters in the study of phase transition of vesicles. The characteristic peculiarity of the concentration increment of sound velocity, $[u]$, is its sharp minimum at the T_{m} (Fig. 4A, curve 1). At temperatures below the T_{m} of DMPC (i.e., $T < 24\ ^{\circ}\mathrm{C}$), the value of $[u]$ slowly decreases with increasing temperature, whereas an increase of $[u]$ is characteristic for higher temperature regions above the T_{m}. This characteristic shape of the plot of $[u]$ versus temperature was described in a number of publications dealing with sound velocimetry studies of phase transitions in vesicle suspensions composed of saturated PCs (see, e.g., Ref. [82]). The nature of the minimum in the plot of $[u]$ versus temperature can be explained as follows. According to Schaaffs [83], the adiabatic compressibility, β_{S} (see Eq. 13), is related to the isothermal compressibility, β_T, and the heat capacities, c_V and c_P, at constant volume and constant pressure:

$$\beta_{\mathrm{S}} = \beta_T c_V / c_P, \tag{13}$$

so that

$$u = \left[c_p/(c_V \rho \beta_T)\right]^{1/2}. \tag{14}$$

Hence, the minimum in the $[u]$ value reflects the effects from both the increasing heat capacity, c_P [84], and isothermal compressibility, β_T, upon approaching the T_{m}. The value β_T is an additional parameter compared to the T_{m} conventionally used in DSC.

According to Eq. (12), in order to calculate the specific apparent compressibility, $\varphi_{\mathrm{K}}/\beta_0$, it is necessary to determine the density of the vesicle solution, ρ, and consequently the specific volume, φ_V. In the case of phospholipids, this parameter changes considerably at the T_{m}. The value of φ_V, as well as the behavior of its plot versus temperature, provides important information about the nature of the lipid phase transition of membranes, for example, degree of cooperativity and expansion coefficient [85]. The typical plot of density, ρ, and specific volume, φ_V, for LUVs composed of DMPC is presented in Fig. 4B, curves 1 and 2, respectively. We can see that density monotonically decreases, while specific volume increases with increasing temperature. Using Eq. (12), the apparent specific compressibility, $\varphi_{\mathrm{K}}/\beta_0$, can be calculated. The plot of $\varphi_{\mathrm{K}}/\beta_0$ as a function of temperature for LUVs composed of DMPC is presented in Fig. 4A, curve 2. We can see that below the phase transition temperature, the value of $\varphi_{\mathrm{K}}/\beta_0$ monotonically increases with temperature, reaches a maximum at

T_m, and then decreases at $T > 24$ °C. This is in agreement with previously reports [39] and can be explained as follows. The apparent specific compressibility can be expressed as

$$\varphi_K/\beta_0 = \varphi_V + [\beta_S], \tag{15}$$

where $[\beta_S] = (\beta_S - \beta_{S0})/(\beta_{S0}c)$ is the concentration increment of the adiabatic compressibility (see Ref. [39]). Thus, the increase of the φ_K/β_0 value in the phase transition region is related to both an increasing specific volume, φ_V, and an increase of β_S. Quite remarkably, φ_K/β_0 decreases when going from 25 °C, slightly above T_m, to 30 °C. In this temperature range φ_V increases (see Fig. 4B, curve 2). We therefore conclude that the decreasing φ_K/β_0 values result from a decrease in β_S. The heat capacity, c_p, also decreases in this temperature range. According to Eq. (13), the temperature dependence in the φ_K/β_0 above T_m thus seems to reflect the isothermal compressibility of lipid bilayers, which is related to the volume fluctuations of the samples and which decreases with temperature [39].

$$\beta_T = \frac{1}{V}\left(\frac{\partial V}{\partial p}\right)_T = \frac{\bar{V^2} - \bar{V}^2}{\bar{V}RT}. \tag{16}$$

Here, R denotes the gas constant [40].

The phase transition temperature of LUV and MLV is similar; however, changes of thermodynamic and mechanical parameters of MLV at the phase transition region are sharper, which evidence on higher cooperativity of transitions in MLV. Using the ultrasound velocimetry and densitometry methods, we determined the changes of $[u]$ and φ_V, values of MLV of DMPC containing various amount of GS as a function of temperature, and using Eq. (12) determined the specific apparent adiabatic compressibility, φ_K/β_0 [40]. The plot of φ_K/β_0 as a function of temperature for MLVs of different GS content is presented in Fig. 5. The shape of the curves are similar for relatively low GS content (below 4 mol%) and are in agreement with previously reported results for pure DMPC unilamellar vesicles [39]. The physical nature of the shape of the plot φ_K/β_0 versus temperature for unmodified DMPC vesicles was discussed earlier. The shape of this dependence is similar for vesicles modified by GS; however, a more pronounced decrease of the adiabatic compressibility in the high-temperature region of the 10 mol% GS sample is evident. From Fig. 5, there is also evidence for increased adiabatic compressibility at relatively low GS concentrations (below 10 mol%) since the adiabatic compressibility in the low-temperature region is still higher than that for pure DMPC; however, it does not significantly differ from that for 2 mol% GS. In the high-temperature region ($T > 25$ °C), adiabatic compressibility for samples containing 10 mol% of

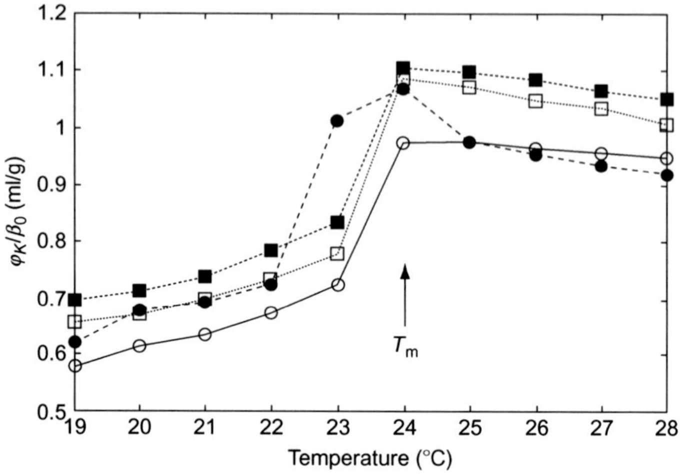

Figure 5 Plot of apparent specific partial compressibility φ_K/β_0 versus temperature for aqueous suspension of MLVs of DMPC with different lipid/peptide molar ratios: 1:0 (○), 1:50 (□), 1:25 (■), and 1:10 (●). The arrow indicates the phase transition temperature (according to Krivanek *et al.* [40] with permission of Elsevier).

GS is even lower compared to pure DMPC. This effect might be caused by the ability of GS to disturb the integrity of the lamellar phase of the lipid bilayer at higher peptide concentrations [86].

The effect of GS on lipid ordering was a matter of debate in the literature for some time. Zidovetzki *et al.* [87] utilized ^{2}H spectroscopy to investigate the effect of GS on the hydrocarbon chain orientational order profiles of chain-deuterated DMPC MLVs in the gel and liquid-crystalline states. They reported no major changes in the gel state, suggesting that GS is excluded from the gel phase. However, above the T_m, the peptide appeared to disorder the hydrocarbon chains. The FTIR data [80] clearly show a partial penetration of GS into the gel phase of DMPC. Susi *et al.* [88], using Raman spectroscopy, reported that GS disordered the hydrocarbon chains below the broadened and reduced gel to liquid-crystalline phase transition but had no effect at temperatures above the phase transition. In contrast, Katsu *et al.* [89], using DPH fluorescence polarization spectroscopy, found no effect above or below the T_m, whereas Yagi *et al.* [90], using the same technique, reported a slight disordering of the liquid-crystalline phase. In contrast, Mihailescu and Horvath [91], using ESR spectroscopy, reported the existence of about six molecules of motionally restricted phospholipids (DMPC or dimyristoylphosphatidylserine-DMPS) associated with each molecule of GS between 30 and 40 °C. The DSC data [10] show a destabilizing effect of GS on the gel phase of DMPC as indicated by the concentration dependent decrease of T_m.

The results obtained in our work demonstrate that GS at contents up to 10 mol% increases the adiabatic compressibility of vesicles. The membrane compressibility is closely related to its ordering and to changes in the hydration of lipid bilayer. This has been shown previously on various planar and spherical membrane structures of different lipid compositions [5,39,80]. We should, however, note that in our study we prepared liposomes together with peptide, that is, we studied exclusively the thermodynamic behavior of lipid bilayers with already incorporated peptide. Therefore, we suppose that the changes of hydration had no significant effect on membrane thermodynamics in comparison with the effect of peptide on the structural properties of lipid bilayer. We can therefore conclude that GS induces a disordering in both the gel and liquid-crystalline state of the lipid bilayer. The ability of GS to increase the membrane compressibility and to decrease the phase transition temperature is evidence for regions of distorted membrane structure around GS molecules.

The ultrasound velocimetry and densitometry have been applied also by the study of the interaction of short α-helical peptides with lipid bilayers of LUV composed of phospholipids of various lengths of hydrocarbon chains. This allowed us to analyze the effect of mismatch between the protein and the lipid on the mechanical properties of the membrane [41].

Earlier, it has been shown that changes in the thermodynamic properties of lipid bilayers containing peptide P_{24}, the structural analogue of L_{24}, can be explained by a mismatch between the hydrophobic part of the peptide (composed of 24 leucine residues) and hydrophobic part of the lipid bilayer [63]. As revealed from X-ray diffraction studies of the P_{24} in PC membranes, the peptide is oriented perpendicular to the lipid bilayer [60]. The same conclusion holds also from NMR studies of the α-helical peptides of comparable length and of hydrophobic core composed of leucine or altered leucine–alanine residues [92]. It is undoubtedly that the presence of polar lysine residues at both ends of L_{24} facilitates anchoring of the peptide in the membrane and prevents the peptides from aggregation due to repulsive electrostatic forces. However, the lysine residues do not affect the conformation of the polar head groups of phospholipids as revealed from FTIR [93] and NMR studies [91]. Certain influence on phospholipid head groups has been observed only for too short (<12 carbons) or too long (>20 carbons) PC [92]. Moreover, the FTIR studies of the C=O stretching band of the phospholipids revealed that it has not been affected by P_{24} or phase state of the phospholipid. It has been suggested that P_{24} probably does not interact significantly with the polar/apolar region of PC bilayers [93]. Therefore, the study of the influence of the P_{24} or L_{24} peptides on physical properties of lipid bilayers was focused on the analysis of hydrophobic interactions. Based on molecular modeling studies, it has been estimated that the hydrophobic length of the α-helix composed of 24 leucine residues is approximately 3.1 nm [63,66]. (Note that this is lower in comparison with routinely calculated end-to-end distance of this α-helical peptide, assuming 0.15 nm projection on a *z*-axis per amino acid residue [58].) Therefore, by using PCs with different lengths of the hydrocarbon chains, it is possible to analyze the influence of L_{24} on thermodynamic and mechanical properties of lipid bilayer depending on the hydrophobic thickness of the membrane. In analogy with a previous study focused on the application of DSC and FTIR methods [63], we applied ultrasound velocimetry and densimetry to study the physical properties of LUVs composed of 13:0, 14:0, 15:0, or 16:0 PCs without and with L_{24} [41]. These saturated phospholipids differ with length of their hydrophobic chains, which resulted in different hydrophobic thickness of lipid bilayer, both in the gel and in liquid-crystalline states, as well as in different temperatures of the main transition. The temperatures of the main transition, as well as hydrophobic thickness of these phospholipids, are shown in Table 1. As can be seen from this table, the temperature of the main phase transition from gel to liquid-crystalline state of the bilayer increases with increasing the number of carbon atoms in the lipid hydrocarbon chain. This is a well-known phenomenon, which is due to increase of the van der Waals and hydrophobic attractive forces between the chains with increasing of their length. We can also see that for all PCs listed in Table 1, the hydrophobic length of the peptide is lower than that for hydrophobic thickness of the bilayer in the gel state, but longer than hydrophobic thickness of bilayer in its liquid-crystalline state. Considering this mismatch between the hydrophobic length of L_{24} and the hydrophobic thickness of the lipid bilayer, it is interesting to analyze how this mismatch will affect the physical properties of the lipid bilayers.

The dependences of [*u*] value versus temperature for suspensions of LUVs without and with L_{24} at mole ratio PC:L_{24} = 100:1 for all PCs studied revealed that for the pure PCs, in all cases the plot of [*u*] versus temperature has the typical shape with minimum at the temperature of phase transition of the respective phospholipid and that the shift of the

Table 1 Temperature of the main phase transition and hydrophobic thickness of the bilayer formed by selected saturated phosphatidylcholines (PCs)

PC	T_m[a] (°C)	Hydrophobic thickness[a] (nm): Gel phase	Liquid-crystalline phase	Mean[b]
13:0	12.8	3.15	2.10	2.63
14:0	23.7	3.42	2.28	2.85
15:0	34.0	3.68	2.45	3.07
16:0	41.5	3.94	2.63	3.29

[a] Adopted from Refs. [63,94].

[b] The mean value of the hydrophobic thickness of the gel and liquid-crystalline phases.

minimum of $[u]$ toward lower temperature exists for LUVs contained L_{24} except 15:0 PC, for which also the differences between $[u]$ values for LUVs, without and with L_{24} are minimal in comparison with other PCs. This peculiarity probably reflects the fact that the mean hydrophobic thickness of 15:0 PC almost matches the hydrophobic length of L_{24}. This result agrees well with that obtained earlier by DSC method on the structurally related peptide P_{24} with an identical hydrophobic length to L_{24} [63]. The determination of specific volume indicated that for LUVs composed of 14:0 and 16:0 PCs, the φ_V values for LUVs containing peptide are higher than those for LUVs without peptide. Finally, for LUVs composed of 15:0 PCs, the differences between values of φ_V for LUVs with and without peptide are not significant in either the gel or liquid-crystalline states. Having both $[u]$ and φ_V values, we analyzed the effect of peptide on apparent specific compressibility of liposomes. In general, the apparent specific compressibility of LUVs contained peptide in a mole ratio PC:L_{24} = 100:1 was higher for pure LUVs in both gel and liquid-crystalline states except for 15:0 PC, for which the differences are small in the gel state and negligible in a liquid-crystalline state. Thus, on the basis of these results, we can conclude that L_{24} increases the compressibility of both LUVs and MLVs in the gel state of lipid bilayers for all PCs studied. The influence of L_{24} on the compressibility of vesicles in the liquid-crystalline state is not so clearly expressed, but there is certainly a tendency to decrease the compressibility for higher peptide concentration.

The thermodynamical properties of the vesicles with incorporated peptide derived from the study of the specific volume were similar to that reported earlier on the basis of the application of precise DSC to study of the influence of the structurally related peptide P_{24} on the properties of bilayers of various hydrophobic thicknesses [63]. Namely, in both studies, the decrease of overall main phase transition temperature and increase of width of phase transition were observed. The decrease of the phase transition, as well as decrease of the cooperativity of the gel to liquid-crystalline phase transition, have been attributed to indirect effect of the presence of the peptide in a lipid bilayer that act as certain impurities (see also Ref. [95] for discussion of this phenomenon). In fact, as we mentioned in Sect. 1, the so-called mattress model of protein–lipid interaction proposed by Mouritsen and Bloom [4] predicts the dependence of the shift of phase transition temperature, ΔT_m, upward or downward of the main transition of pure phospholipid depending on the degree of mismatch between the hydrophobic length of peptide (h_p) and the hydrophobic thickness of lipid bilayer in a gel (h^g) and in a fluid (h^f) state as well as on the mole ratio between peptide and phospholipid, x_p (see Eq. 1). Thus, according to Eq. (1) and using the hydrophobic length of L_{24} and membrane hydrophobic thickness in the gel and liquid-crystalline states (see Table 1), we could expect the shift of ΔT_m upward relatively to the main transition for 13:0 and 14:0 PCs, almost no effect for 15:0 PC and shift downward for 16:0 PC. Therefore, when hydrophobic length of the peptide is equal to the mean hydrophobic thickness of the bilayer, no shift of the T_m is observed. As a matter of fact, using precise DSC method, it was possible to analyze the influence of peptide P_{24} on the main transition for PCs of various lengths of saturated hydrophobic chains. It has been shown that the behavior of the peptide containing bilayers does not approximate that of ideal "two-dimensional solution" and that the peptide induces a two-component DSC of the membrane. The sharp component with transition temperature T_f has been attributed to a bulk PC (peptide-poor domains), while broad component with transition temperature T_b corresponded to peptide-associated phospholipids (peptide-rich domains). The plot $\Delta T_m = T_b - T_f$ versus mean hydrophobic thickness of lipid bilayer was in good agreement with the mattress model, that is, ΔT_m was shifted upward for PC with $C < 15$ (C is the number of carbon atoms in hydrophobic chain of PC) and downward for $C > 15$. $\Delta T_m = 0$ for $C = 15$ (see Ref. [63] for more details). As a matter of fact, this has been confirmed also in our experiments. Certainly, the nonzero shift of T_m was observed for all PCs except 15:0, for which ΔT_m was practically 0. For bilayers composed of 15:0, there were minimal differences between the values of $[u]$, φ_V, and φ_K/β_0 without and with L_{24}.

The apparent specific compressibility of lipid bilayer, φ_K/β_0, can be used for estimation of the influence of the peptide on surrounding lipid bilayer. The plot of the changes of φ_K/β_0 as a function of L_{24}/DMPC mole ratio for a gel ($T = 16$ °C) and liquid-crystalline state ($T = 32$ °C) of lipid bilayer is presented on Fig. 6A. We can see that in a gel state, a remarkable increase of the φ_K/β_0 takes place and saturation occurs already at a mole ratio of 0.02, while in a liquid-crystalline state only minor changes of φ_K/β_0 were observed. Thus, the L_{24} peptides in the gel state of the membrane are able to change the structural properties of the lipid bilayer that involve around 50 lipid molecules. This number is two times larger than estimation based on DSC studies [63] and corresponds to approximately two "solvation layers" of phospholipids around peptide. This higher value may be connected with the peculiarities of ultrasound velocity methods, which reflects the changes of the physical properties of the membrane as a whole. On the other hand, the value of φ_K/β_0 for MLV at higher temperatures ($T > 30$ °C) when the bilayer is in a liquid-crystalline state practically does not depend on the content of the peptide. This suggests that in a liquid-crystalline state, which is natural conditions of the cell membranes, the peptide is not able to change mechanical properties of the membrane in a longer distance than those corresponding to maximum one layer of lipids surrounding the peptide. As revealed from Fig. 6A, the maximal changes of $\Delta\varphi_K/\beta_0 = 0.07$. This is much lower than the changes of φ_K/β_0 due to the phase transition of the bilayer from gel (G) to liquid-crystalline (LC) state: $\Delta\varphi_K/\beta_0 = (\varphi_K/\beta_0)_{LC} - (\varphi_K/\beta_0)_G = 0.47$ ml/g.

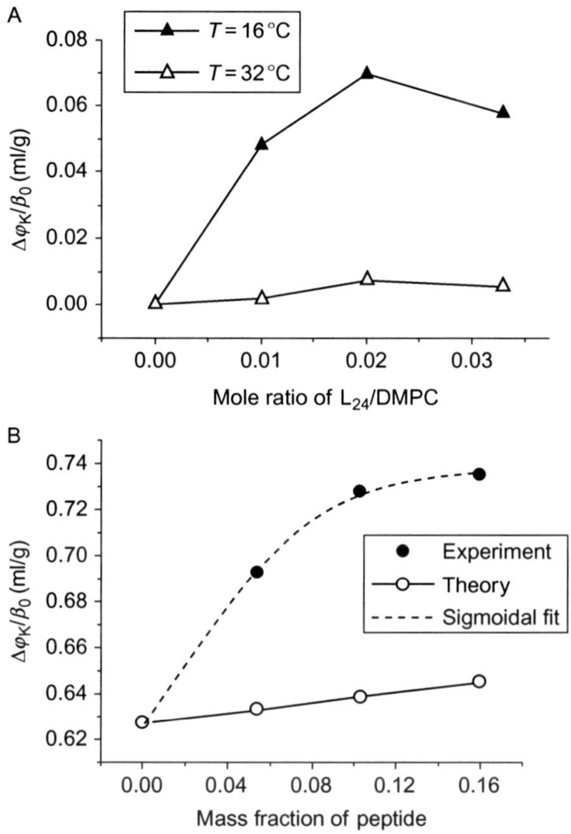

Figure 6 (A) Plot of the changes of normalized specific apparent compressibility ($\Delta\varphi_K/\beta_0 = \varphi_K/\beta_0 - (\varphi_K/\beta_0)_0$, where $(\varphi_K/\beta_0)_0$ is the apparent compressibility of vesicles without peptide and φ_K/β_0 those at certain peptide/DMPC mole ratio), as a function of the mole ratio of L_{24}/DMPC in MLV at $T = 16$ °C (▲) and 32 °C (Δ), respectively. See legend at the figure. (B) Plot of the apparent specific compressibility as a function of the mass fraction of the peptide (x) in a gel state of lipid bilayer ($T = 16$ °C) obtained experimentally (●) and using Eq. (18) (○). The dashed line is a sigmoidal fit of experimentally obtained data (see legend) (according to Rybár *et al.* [41] with permission of Elsevier).

The question arises about the nature of the changes of mechanical properties of the lipid bilayer induced by the L_{24}. At least, two types of processes could be responsible for this behavior:

1. Incorporation of the peptide disturbs the membrane and this causes the changes of the mechanical properties, or incorporation of the peptide affects the mechanical properties of the membrane due to contribution from its own compressibility, or both the peptide compressibility and its

disturbing effect are responsible for the changes of the membrane compressibility.
2. Changes of compressibility are due to changes of vesicle hydration caused by incorporation of the peptides.

Let us discuss these possibilities.

The apparent molar compressibility of the system, ΦK_S ($\Phi K_S = \beta_0 M(\varphi_K/\beta_0)_S$, where M is the molecular weight of the solute), involves the intrinsic compressibility of the lipid bilayer, K_L, the peptide K_P, and the compressibility effect of hydration, ΔK_h:

$$\Phi K_S = m_L K_L + m_P K_P + n_h (K_{Sh} - K_{S0}), \tag{17}$$

where m_L and m_P are the mole fractions of lipids and proteins, respectively ($m_L + m_P = 1$); n_h is the number of water molecules in hydration shell; and K_{Sh} and K_{S0} are the partial molar compressibilities of the water in hydration shell and of the solvent (bulk water), respectively [96]. Equation (17) describes the apparent specific compressibility of the vesicles and their components relatively to the distilled water that was used as a solvent. Because the solvent in the vesicle interior is identical with that outside the vesicles, its effect on the hydration of the vesicles is symmetrical.

Let us first discuss the possible contribution of the peptide L_{24} to the compressibility of the vesicles. The substantial part of the peptide is located in the hydrophobic part of the membrane. Two positively charged lysines at each peptide side, however, contact with the surrounding water. We can therefore expect the contribution to the peptide compressibility from both the compressibility of its hydrophobic part and the hydrated shell surrounding lysines. Our results show that the adiabatic compressibilities of the vesicles are different in a gel and in a liquid-crystalline state. While in a gel state the compressibility depends on the concentration of peptide, in a liquid/crystalline state and at temperature 32 °C, the compressibility of MLV does not depend on the content of the peptide. However, in the case of LUV, the compressibility of vesicles depends on the peptide content also at higher temperatures (30 °C). On the other hand, this dependence is nonmonotonous, that is, the compressibility of LUV with higher peptide content is lesser in comparison with lower one. In this respect, the behavior of MLV and LUV is similar at the temperatures below 30 °C. As it has been shown by FTIR method, the peptide of similar structure to L_{24}, that is, P_{24} is rather stable and its structure practically does not change with temperature in the temperature range studied in this work. Also, the phase transition from gel to liquid-crystalline state practically does not affect the peptide structure. We can therefore assume that also the compressibility of the peptide will not substantially change with temperature at the range of 16–32 °C and following phase transition. To discuss the effect of the compressibility of the peptide into the overall compressibility of vesicles, it is desirable to estimate the value of apparent adiabatic compressibility of the peptide, $(\varphi_K/\beta_0)_P$. For this purpose, we should present the apparent adiabatic compressibility as follows:

$$\varphi_K/\beta_0 = x(\varphi_K/\beta_0)_P + (1-x)(\varphi_K/\beta_0)_L, \qquad (18)$$

where x is the mass fraction of the peptide in a peptide–lipid mixture ($x = x_P/(x_P + x_L)$, where x_P is the mass of the peptide and x_L that of the lipid). The plot of the value φ_K/β_0 as a function of mass fraction of peptide and for a gel state of the bilayer is shown on Fig. 6B. This plot can be well fitted by sigmoidal function presented by dashed line. According to this fit, at $x = 1$, we obtain for $(\varphi_K/\beta_0)_P = 0.74$. This is rather high value in comparison with that for a lipid and with previously reported values for α-helical polypeptides and globular proteins in water [97]. Certainly, the apparent specific compressibility of α-helix of polylysine in water is due to high degree of hydration negative: −0.73 ml/g. The adiabatic compressibility of globular protein human serum albumin is positive due to more compact structure that is less exposed to a water and reaches the value of 0.13 ml/g (see also Refs. [98,99]). We should note that so far, the adiabatic compressibility of peptide in a membrane has not been determined. Considering the fact that most part of the peptide is in hydrophobic interior of the membrane and only polar part is exposed to a water and thus hydrated, we can expect that overall specific compressibility of the peptide should be positive in analogy with globular proteins. However, taking into account the structural stability of the peptide with temperature, we can also expect that this should be rigid structure, at least not softer than the core of the globular proteins. Therefore, the value obtained above, that is, $(\varphi_K/\beta_0)_P = 0.74$ is likely overestimated. We assume that $(\varphi_K/\beta_0)_P$ should be similar or lower than that of globular proteins, that is, lower in comparison with compressibility of lipid bilayer. Unfortunately, measurement of specific compressibility of liposomes with higher peptide content is technically complicated. However, let us use the value $(\varphi_K/\beta_0)_P$ obtained by fitting and construct the theoretical dependence φ_K/β_0 as a function of mass fraction of peptide, x, using Eq. (18) and known value of φ_K/β, for pure lipid bilayer $(\varphi_K/\beta_0)_L = 0.63$ ml/g. This plot is presented on Fig. 6B. We can see that the values calculated according to Eq. (18) are much smaller in comparison with that obtained experimentally. Obviously, for smaller adiabatic compressibility of peptide $(\varphi_K/\beta_0)_L$, the differences between experimentally and theoretically obtained dependences should be higher. Therefore, we assume that the increase of apparent adiabatic compressibility of the vesicles with increased concentration of peptide is caused by disturbing effect of the peptide on the lipid bilayer and not by the compressibility of the peptide itself.

The analysis of the effect of hydration [41] showed that it has negligible effect on the adiabatic compressibility of the vesicles with incorporated peptides and is not higher than 2%. We can therefore conclude that the main contribution to the changes of adiabatic compressibility of vesicles is due to the disturbing effect of the peptide on surrounding hydrophobic chains of the phospholipids.

Figure 6A shows also rather interesting phenomena consisting in non-monotonous changes of the adiabatic compressibility at higher peptide concentration. This is especially well expressed for a gel state of the bilayer. We assume that this may be connected with influence of distorted lipid regions around peptide to each other. Similar phenomenon was observed earlier on the planar BLMs contained BR. The elasticity modulus of these membranes increased during illumination of the membrane, that is, due to conformational changes of bacteriorhodopsin. However, after reaching certain saturation, the elasticity modulus started to decrease [13]. Thus, at higher peptide concentration, not only the peptide affects the ordering and physical properties of the lipid bilayer, but also the distorted lipid layers around the peptide interact and thus contribute to physical properties of the membrane.

Thus, we showed that the model integral peptide induces an increase of the compressibility of lipid bilayer of vesicles in a gel state. This is due to disturbing effect of the peptide on the ordering of hydrocarbon chains of the lipids. We suspect that around the peptide, a disturbing part of bilayer arises that involves at least two molecular layers of phospholipids. The peptide, however, affects only weakly the compressibility of the membrane in a liquid-crystalline state.

3.2. Effect of Na,K-ATPase on the Mechanical Properties of the Lipid Bilayers

In contrast with short peptides [100], the mechanism of interaction between the Na,K-ATPase and phospholipids has been studied to a lesser extent. The focus was concentrated on how phospholipid structure and composition affect the functional properties of the protein, including transport rate, ion binding, and turnover. A higher transport rate was observed when the ion pumps were reconstituted in membranes of di(18:1)PC, di(20:1)PC, and di(22:1) PC, whereas in di(14:1)PC and di(16:1)PC, the enzyme was almost inactive [101,102]. In certain cases also specific lipids have been found to be an important factor for functioning the integral protein. For example, Na,K-ATPase interacts specifically with cardiolipin [103]. ATPase activity was considerably enhanced in the presence of relatively high amount of cholesterol (40 mol%) [104]. In this section, we report the results of investigation of the effect of Na,K-ATPase on the specific volume and compressibility of lipid bilayers by means of the ultrasound velocimetry and densitometry methods.

Na,K-ATPase was prepared from outer medulla of rabbit kidneys using procedure C described in Ref. [105]. The specific activity of the purified microsomal preparation was ~2200 μmol P_i per hour and mg protein at 37 °C. Unilamellar liposomes were prepared from cholate-solubilized protein and lipids (dioleoylphosphatidylcholine-DOPC) by dialysis method [106] with various protein/lipid molar ratios.

Based on ultrasound velocimetry and densitometry, we determined specific volume, φ_V, and the specific apparent compressibility, φ_K/β_0, of proteoliposomes as a function of molar ratio of ATPase/DOPC. These values are presented in Table 2. It can be seen from this table that with increased content of ATPase, both specific volume and specific compressibility decrease. The lower specific volume of proteoliposomes in comparison with unmodified membranes may indicate a more compact structure of lipid bilayer in the presence of the ATPase. The behavior of specific compressibility reveals a higher ordering of the lipid bilayer with increasing contents of protein. However, also another factors such as compressibility of protein itself and changes in hydration could contribute to the compressibility changes. Therefore, all aforementioned factors should be carefully analyzed.

The concentration increment of sound velocity as well as the specific volume, and consequently, also the compressibility are additive and integral parameters. This means that if, for example, a protein molecule is incorporated into the lipid bilayer, the overall specific apparent compressibility, φ_K/β_0, will be composed of the sum of the three respective components of the membrane, that is, (1) of the undisturbed lipid bilayer, $(\varphi_K/\beta_0)_L$, (2) of the altered part of the bilayer caused by protein–lipid interactions, $(\varphi_K/\beta_0)_{LP}$, and (3) of the protein molecule itself, $(\varphi_K/\beta_0)_P$. In addition, the contribution of the hydration shell, $(\varphi_K/\beta_0)_H$, to the overall compressibility should also be taken into account. Therefore, the overall apparent specific compressibility of proteoliposomes is given by the equation:

Table 2 Properties of the lipid bilayers with incorporated Na,K-ATPase calculated on the basis of Eq. (21)

Molar ratio ATPase:lipid	φ_V (ml/g)	φ_K/β_0 (ml/g)	s_L	s_P	N_V	N_β
0:1	0.999	1.053	1	0	–	–
1:52,500	0.980	1.003	0.997	0.0027	23,700	8000
1:5250	0.954	0.942	0.973	0.0272	5120	290

s_L and s_P is the fraction of the area occupied by lipid and protein, respectively. N_V and N_β are numbers of lipid molecules per one molecule ATPase affected by protein calculated according to the specific volume and intrinsic compressibility changes. The values of φ_V and φ_K/β_0 were determined at $T = 25$ °C.

$$\varphi_K/\beta_0 = (\varphi_K/\beta_0)_L + (\varphi_K/\beta_0)_{LP} + (\varphi_K/\beta_0)_P + (\varphi_K/\beta_0)_H. \tag{19}$$

The hydration term is negative at relatively low temperatures, while the other terms are positive. Therefore, the overall value of the specific apparent compressibility could be positive or negative depending on the degree of hydration and on temperature. For example, at 20 °C, the compressibility of globular proteins is in the range 0.05–0.15 ml/g, while for more hydrated fibrillar proteins the values of apparent compressibility are in the range −0.1 to 0.75 ml/g [97]. The value of the overall apparent compressibility is therefore a sensitive indicator of the structural and conformational state of the protein. We will therefore analyze the possible contribution of aforementioned terms to the compressibility of proteoliposomes. First, we discuss the effect related to the decreasing compressibility of proteoliposomes with the increasing ATPase concentration in the membrane.

We have shown that the coefficient of reduced apparent specific compressibility decreases with an increasing ATPase/DOPC ratio. This effect can be related to several processes:

1. Due to considerably smaller compressibility of proteins in comparison with the compressibility of a lipid bilayer, the decrease of the compressibility will be caused by an increasing number of protein molecules in a membrane, arising from the additive rule mentioned earlier.
2. Since the ATPase is composed not only from a hydrophobic membrane domain, but also from hydrophilic, cytosolic part exposed to the electrolyte, the latter is hydrated, and the decrease of the overall compressibility can possibly be related also to an increased hydration.
3. The ATPase strongly interacts with surrounding lipids, and as a consequence, the ordering of the lipid bilayer increases.

The aforementioned possibilities have to be estimated and discussed.

Effect of the compressibility of the protein. The adiabatic compressibility of proteins depends on the degree of their hydration. For example, at 20 °C, the compressibility of human serum albumin (globular protein) is approximately 0.13 ml/g, while the compressibility of fibrillar proteins, for example, rabbit myosin, is negative: −0.2 ml/g [97]. In our recent study, we showed that the reduced apparent specific compressibility of native sarcoplasmic reticulum isolated from rabbit skeletal muscles which contained Ca-ATPase was about −0.46 ml/g at 25 °C [107]. Our results evidence that the value of the apparent adiabatic compressibility for liposomes with even the highest amount of Na,K-ATPase (molar ratio ATPase/DOPC = 1:5250) was positive over the whole temperature range studied (10–38 °C) [108]. This suggests that the structure of the Na,K-ATPase is rather compact and that the surface of hydration shell is lower compared to Ca-ATPase. Let us assume that the compressibility of the liposomes is composed of two parts:

lipid bilayer and the protein. Then, the overall normalized apparent specific compressibility, φ_K/β_0, will be given by equation [5]:

$$\varphi_K/\beta_0 = s(\varphi_K/\beta_0)_L + (1 - s)(\varphi_K/\beta_0)_P, \tag{20}$$

where $(\varphi_K/\beta_0)_L$ and $(\varphi_K/\beta_0)_P$ are the normalized apparent specific compressibilities of the lipid bilayer and the protein, respectively; s is the relative area occupied by the lipid bilayer; and $1 - s$ that of the protein. The cross-sectional area of Na,K-ATPase estimated on the basis of electron microscopy is approximately 48.7 nm^2 [109]. The area, S_L, occupied by one phospholipid molecule in a liquid-crystalline state is approximately 0.7 nm^2 [110]. Then, for the molar ratio ATPase/DOPC = 1:5250, the relative area occupied by lipids will be $s = 1 - 2mS_P/S_L = 1 - 2 \times (1/5250)(48.7/0.7) = 0.973$, where m is the molar fraction of ATPase:lipid; S_L and S_P are the area occupied by lipid and protein, respectively; and the coefficient 2 is due to the pair of the phospholipids in a bilayer. The relative area occupied by protein is $1 - s = 0.027$. Using Eq. (20) and the measured value of $(\varphi_K/\beta_0)_L = 1.12$ ml/g (at $T = 38$ °C [108]), it is possible to estimate the apparent adiabatic compressibility of the protein; applying a value of $(\varphi_K/\beta_0) = 0.986$ ml/g at this temperature, one obtains from Eq. (20) $(\varphi_K/\beta_0)_P = -3.84$ ml/g. Thus, the compressibility of the protein could explain the decrease of the overall compressibility of the liposomes with an increased number of protein molecules. The estimated value of $(\varphi_K/\beta_0)_P$ is, however, negative and considerably lower than the compressibility of the fibrillar and globular proteins investigated so far [97]. The Na,K-ATPase is composed of two subunits. The α-subunit (~113 kDa) binds ATP, and sodium and potassium ions, and it contains the phosphorylation site. The smaller β-subunit, a glycoprotein (~35 kDa) has yet unknown functional properties but it is necessary for enzyme functioning [111]. The substantial part of the Na,K-ATPase is anchored in hydrophobic part of the membrane. Therefore, it could be expected that the value $(\varphi_K/\beta_0)_P$ should be positive like that for globular proteins [97,98]. Thus, it seems to be unlikely that the observed changes of apparent specific compressibility are related exclusively to that of the proteins, but rather to the compressibility of proteolipid complexes.

Effect of hydration. The decrease of the reduced apparent specific compressibility with an increasing concentration of the Na,K-ATPase could be related to an increase of the hydration of the liposomes. However, this effect was observed not only at relatively low temperatures at which the hydration shell is highly ordered (and its compressibility is lower in comparison with unbounded water molecules), but also at higher temperatures ($T = 38$ °C), at which the compressibility of the hydration shell is comparable with the compressibility of surrounding water molecules. This is true at least for zwitterionic phospholipids such as DPPC or DOPC [14]. We assume, therefore, that a hydration effect is not predominant in the observed changes of the apparent specific compressibility of the proteoliposomes.

Effect of protein–lipid interactions. The changes of the reduced apparent specific compressibility of the liposomes (0.093–0.14 ml/g), which were induced by the Na,K-ATPase at its highest concentration (molar ratio ATPase/DOPC = 1:5250), are comparable with changes of that of liposomes during the phase transition of phospholipids [19,112]. Therefore, the observed changes could be explained by an influence of the enzyme on the structural state of the surrounding phospholipid molecules. Probably, due to interactions between the hydrophobic side of the membrane domain of the Na,K-ATPase and the hydrocarbon chains of the phospholipids, the lipid bilayer becomes more ordered, which is reflected by the decrease of compressibility.

Let us make estimations of the number of lipid molecules whose structural state may be affected by ATPase. This estimation can be based on either the changes in specific volume or apparent specific compressibility. First, we consider the changes of specific volume. As it is seen from Table 2, the specific volume of the lipids decreases with increased ATPase concentrations. Let Q be any specific physical property (normalized per mass) of a flat lipid membrane containing protein (i.e., specific volume or compressibility). Then, in analogy to Eq. (20), the following equation is approximately valid if the protein has the same thickness as the lipid bilayer:

$$Q = sQ_L + (1 - s)Q_P, \tag{21}$$

where s and $(1 - s)$ is the surface fraction occupied by lipid and protein, respectively; Q_L is the average property of the lipid affected by protein; and Q_P is the property of the protein. The number of lipid molecules affected by ATPase can be calculated on the basis of specific volume changes as follows. Let Q be the specific volume, then according to Eq. (21), the average specific volume of the lipid affected by protein is $Q_L = [Q - (1 - s)Q_P]/s$, where $Q_P \approx 0.73$ ml/g is the specific volume for most proteins [113]. Thus, corresponding Q_L values are 0.981 ml/g and 0.960 ml/g for molar ratios of ATPase:lipid 1:52,500 and 1:5250, respectively. These values are less than those for pure lipid (0.999 ml/g). Therefore, the average volume of lipid is reduced by 1.8% and 3.9%, respectively, for above given molar ratios of ATPase:lipid. These changes are comparable with that caused by phase transition of lipids from fluid to solid state which are approximately 4% [110]. If we assume that a portion of lipids indeed undergoes transition to solid state, then we can calculate the number of such lipid molecules $N_V = 52{,}500 \times 1.8/4 = 23{,}700$ and $N_V = 5250 \times 3.9/4 = 5120$ for molar ratios ATPase:lipid 1:52,500 and 1:5250, respectively (see Table 2).

The coefficient 4 corresponds to the aforementioned maximal changes of the specific volume of phospholipids at phase transition. If we assume, similarly to the findings for the Ca-ATPase, that only one layer of annular phospholipids exists around the Na,K-ATPase, then approximately 79 annular lipids surround the protein in a lipid bilayer, assuming a cross-sectional area of 48.7 nm^2 [109]. Thus, the obtained values exceed 64–300 times the number of the closest neighbors of a protein molecule.

Similar calculations can be performed also based on the changes of specific apparent compressibility. The intrinsic compressibility of most of the proteins is in the range of 0.2–0.55 ml/g [97,98]. Considering that the relative changes of intrinsic compressibility of lipid bilayer are $\Delta\beta/\beta = -0.30$ at the transition from fluid to solid state [39,112] and performing calculations in analogy to that for specific volume, the average number of lipid molecules that undergo a transition to the solid state due to the presence of ATPase will be $N_\beta = 8000$ and 290 for molar ratios ATPase: lipid 1:52,500 and 1:5250, respectively (see Table 2). Thus, the obtained values exceed 3.7–100 times of the number of annular lipids around one ATPase molecule. These analyses indicate that the ATPase affects the physical properties of a substantially higher number of lipid molecules than that of lipid annulus.

This conclusion is supported also by the fact that the applied method to study liposome compressibility is a macroscopic one, that is, it does not reflect local effects or the contribution of small structural changes of the bilayer. Considering that the average diameter of unilamellar liposomes used in this study is about 100 nm [106], approximately 17 molecules of ATPase are incorporated in one liposome at a molar ratio ATPase/DOPC = 1:5250, and the proteolipid complexes composed of an ATPase and its annular lipids occupy only 4% of the whole liposome surface. So far performed studies indicate, however, that the detected changes of mechanical properties of the membranes were significant only when at least 10% of the membrane area were occupied by an altered structure [5]. Moreover, in this study, the changes of apparent specific compressibility were observed even at rather low ATPase/DOPC ratio, that is, at 1:175,000. In this case, not more than one ATPase molecule per liposome is expected in the average, and the proteolipid complexes represent only 0.2% of the liposome area. Similar results were reported earlier for the proteoliposomes containing bacteriorhodopsin [14] (see Sect. 1). Therefore, the assumption of the existence of altered membrane regions considerably larger than those created by the ATPases and their annular lipids seems to be realistic.

However, aforementioned assumption on the rigidization of DOPC molecules induced by ATPase is in contradiction with the current theories of the phase transition in a lipid membrane. Certainly, if we assume that ATPase induced transition of lipids into the solid state, then the correlation length should be approximately 100 diameters of lipid molecules. Such a long correlation seems to be impossible at the experimental condition used in our work. As a matter of fact, such a great correlation length could be observed only in *immediate vicinity* to the phase transition temperature (which is -20 °C for the DOPC). As soon as temperature increases, the correlation length is sharply reduced and reaches one to two lipid molecules at temperatures several degrees away from the transition point. This behavior of lipid system is well known from both experimental (calorimetry) and theoretical (Monte Carlo simulation) studies published by many authors (see, e.g., Refs. [114,115]). The temperature at which our experiments were performed exceeds the transition point by 30–60 °C.

The question, however, arises why so enormous changes are induced by ATPase and what is the physical mechanisms of this process. Considering the highly accurate determination of specific volume and apparent specific compressibility, we believe that the calculated changes really characterize the system studied. Nevertheless, rather large number molecules of lipids involved in the interaction with Na,K-ATPase are amazing. We are reporting our observation as an interesting experimental phenomenon whose physical mechanism remains unclear for us. It is unlikely that ATPase is able to induce rigidization of DOPC molecules at the temperature substantially surpassing the phase transition of this lipid. Therefore, there should be another physical mechanisms that may explain the observed phenomena. We let this mechanism to be a matter of further studies.

ACKNOWLEDGMENTS

This work was supported by the Slovak Grant Agency (Project No. 1/0794/10) and by Agency for Promotion Research and Development under the contracts No. APVV-0362-07 and LPP-0341-09. I am grateful to professors H.-J. Apell, R.N. Mc Elhaney, and D.P. Kharakoz for valuable discussion and advice.

REFERENCES

[1] T. Gil, J.H. Ipsen, O.G. Mouritsen, M.C. Sabra, M.M. Sperotto, M.J. Zuckermann, Theoretical analysis of protein organization in lipid membranes, Biochim. Biophys. Acta 1376 (1998) 245–266.

[2] A.G. Lee, Lipid–protein interactions in biological membranes: a structural perspective, Biochim. Biophys. Acta 1612 (2003) 1–40.

[3] P. Wiggins, R. Phillips, Membrane–protein interactions in mechanosensitive channels, Biophys. J. 88 (2005) 880–902.

[4] O.G. Mouritsen, M. Bloom, Mattress model of lipid–protein interactions in membranes, Biophys. J. 46 (1984) 141–153.

[5] T. Hianik, V.I. Passechnik, Bilayer Lipid Membranes: Structure and Mechanical Properties, Kluwer Academic Publishers, The Netherlands, 1995.

[6] P.A. Kralchecsky, V.N. Paunov, N.D. Denkov, K. Nagayama, Stress in lipid membranes and interactions between inclusions, J. Chem. Soc. Faraday Trans. 91 (1995) 3415–3432.
[7] J. Peschke, J. Riegler, H. Möhwald, Quantitative analysis of membrane distortions induced by mismatch of protein and lipid hydrophobic thickness, Eur. Biophys. J. 14 (1987) 385–391.
[8] M.M. Sperotto, O.G. Mouritsen, Monte Carlo simulation studies of lipid order parameter profile near integral membrane protein, Biophys. J. 59 (1991) 261–270.
[9] H.W. Huang, Deformation free energy of bilayer membrane and its effect on gramicidin channel lifetime, Biophys. J. 50 (1986) 1061–1070.
[10] P. Helfrich, E. Jacobson, Calculation of deformation energies and conformations in lipid membranes containing gramicidin channels, Biophys. J. 37 (1990) 1075–1084.
[11] V. Pereyra, PASVA 3: an adaptive finite-difference FORTRAN program for first order nonlinear boundary value problems, Lect. Notes Comput. Sci. 76 (1978) 67–88.
[12] T. Hianik, P. Borak, B. Piknova, Long-range interactions in proteoliposomes containing bacteriorhodopsin, Acta Phys. Univ. Comen. 34 (1993) 241–246.
[13] T. Hianik, L. Vozár, Mechanical response of bilayer lipid membranes during bacteriorhodopsin conformational changes, Gen. Physiol. Biophys. 4 (1985) 331–336.
[14] T. Hianik, B. Piknova, V.A. Buckin, V.N. Shestimirov, V.L. Shnyrov, Thermodynamics and volume compressibility of phosphatidylcholine liposomes containing bacteriorhodopsin, Prog. Colloid Surf. Sci. 93 (1993) 150–152.
[15] D. Marsh, Lipid–protein interactions in membranes, FEBS Lett. 268 (1990) 373–376.
[16] D. Marsh, L.I. Horvath, Structure, dynamics and composition of the lipid–protein interface. Perspectives from spin-labelling, Biochim. Biophys. Acta 1376 (1998) 267–296.
[17] M. Rehorek, N.A. Dencher, M.P. Heyn, Long range lipid–protein interactions. Evidence from time-resolved fluorescence depolarization and energy transfer experiments with bacteriorhodopsin–dimyristoyl phosphatidylcholine vesicles, Biochemistry 34 (1985) 5980–5988.
[18] S.A. Sanchez, M.A. Tricerri, G. Ossato, E. Gratton, Lipid packing determines protein–membrane interactions. Challenges for apolipoprotein A–I and high density lipoproteins, Biochim. Biophys. Acta 1798 (2010) 1399–1408.
[19] C. Toyoshima, M. Nakasako, H. Nomura, H. Ogawa, Crystal structure of the calcium pump of sarcoplasmic reticulum at 2.6 A resolution, Nature 405 (2000) 647–655.
[20] A.P. Starling, J.M. East, A.G. Lee, Effects of phospholipid fatty acyl chain length on phosphorylation and dephosphorylation of the Ca^{2+}-ATPase, Biochem. J. 310 (1995) 875–879.
[21] A.M. Ernst, F.X. Contreras, B. Brügger, B.F. Wieland, Determinants of specificity at the protein–lipid interface in membranes, FEBS Lett. 584 (2010) 1713–1720.
[22] H.I. Petrache, D.M. Zuckerman, J.N. Sachs, J.A. Killian, R.E.H. Koeppe II, T.B. Woolf, Hydrophobic matching mechanism investigated by molecular dynamics simulations, Langmuir 18 (2002) 1340–1351.
[23] D.P. Tieleman, L.R. Forest, M.S.P. Samsom, H.J.C. Berendsen, Lipid properties and the orientation of aromatic residues in OmpF, influenza M2, and alamethicin systems: molecular dynamics simulations, Biochemistry 37 (1998) 17554–17561.
[24] D. Wobschall, Bilayer membrane elasticity and dynamic response, J. Colloid Interface Sci. 36 (1971) 385–396.
[25] D. Wobschall, Voltage dependence of bilayer membranes, J. Colloid Interface Sci. 40 (1972) 417–423.

[26] E.A. Evans, Bending resistance and chemically induced moments in membrane bilayers, Biophys. J. 14 (1974) 923–931.
[27] E.A. Evans, S. Simon, Mechanics of bilayer membranes, J. Colloid Interface Sci. 51 (1975) 266–271.
[28] V.I. Passechnik, About the model of elastic bilayer membrane, Biophysics (Moscow) 25 (1980) 265–269.
[29] E. Evans, W. Rawicz, Entropy-driven tension and bending in condensed-fluid membranes, Phys. Rev. Lett. 64 (1990) 2094–2097.
[30] T. Hianik, Biological membranes and membrane mimics, in: P.N. Bartlett (Ed.), Bioelectrochemistry: Fundamentals, Experimental Techniques and Applications, Wiley, Chichester, 2008, pp. 87–156.
[31] V.I. Passechnik, T. Hianik, Frequency measurement of the modulus of elasticity of BLM, Biophysics (Moscow) 22 (1977) 548–549.
[32] T. Hianik, M. Haburcak, K. Lohner, E. Prenner, F. Paltauf, A. Hermetter, Compressibility and density of lipid bilayers composed of polyunsaturated phospholipids and cholesterol, Colloids Surf. A 139 (1998) 189–197.
[33] T. Hianik, Electrostriction of supported lipid membranes and their application in biosensing, in: V.M. Mirsky (Ed.), Ultrathin Electrochemical Chemo- and Biosensors: Technology and Performance, Springer-Verlag, Berlin, 2004.
[34] T. Hianik, A. Labajova, Electrostriction of supported lipid films at presence of cationic surfactants, surfactant-DNA and DNA-Mg^{2+} complexes, Bioelectrochemistry 58 (2002) 97–105.
[35] M. Schwarzott, P. Lasch, D. Baurecht, D. Naumann, U.P. Fringeli, Electric field-induced changes in lipids investigated by modulated excitation FTIR spectroscopy, Biophys. J. 86 (2004) 285–295.
[36] I. Zawisza, X. Bin, J. Lipkowski, Potential-driven structural changes in Langmuir–Blodgett DMPC bilayers determined by in situ spectroelectrochemical PM IRRAS, Langmuir 23 (2007) 5180–5194.
[37] E. Evans, R. Kwok, Mechanical calorimetry of large dimyristoylphosphatidylcholine vesicles in the phase transition region, Biochemistry 21 (1982) 4874–4879.
[38] W. Rawicz, K.C. Olbrich, T. McIntosh, D. Needham, E. Evans, Effect of chain length and unsaturation on elasticity of lipid bilayers, Biophys. J. 79 (2000) 328–339.
[39] S. Halstenberg, T. Heimburg, T. Hianik, U. Kaatze, R. Krivanek, Cholesterol induced variations in the volume and enthalpy fluctuations of lipid bilayers, Biophys. J. 75 (1998) 264–271.
[40] R. Krivanek, P. Rybar, E.J. Prenner, R.N. McElhaney, T. Hianik, Interaction of the antimicrobial peptide gramicidin S with dimyristoylphosphatidylcholine bilayer membranes: a densitometry and sound velocimetry study, Biochim. Biophys. Acta 1510 (2001) 452–463.
[41] P. Rybár, R. Krivánek, T. Samuely, R.N.A.H. Lewis, R.N. McElhaney, T. Hianik, Study of the interaction of an α-helical transmembrane peptide with phosphatidylcholine bilayer membranes by means of densimetry and ultrasound velocimetry, Biochim. Biophys. Acta 1768 (2007) 1466–1478.
[42] E. Evans, R. Skalak, Mechanics and Thermodynamics of Biomembranes, CRC Press, Boca Raton, FL, 1980.
[43] S.B. Hladky, D.A. Haydon, Ion transfer across lipid membranes in the presence of gramicidin A. I. Studies on the unit conductance channel, Biochim. Biophys. Acta 274 (1972) 294–312.
[44] M. Thompson, U.J. Krull, The electroanalytical response of the bilayer lipid membrane to valinomycin: an empirical treatment, Anal. Chim. Acta 141 (1982) 49–56.

[45] P. Vitovic, S. Kresak, R. Naumann, S.M. Schiller, R.N.A.H. Lewis, R. N. McElhaney, et al. The study of the interaction of a model α-helical peptide with lipid bilayers and monolayers, Bioelectrochemistry 63 (2004) 169–176.
[46] D.P. Nikolelis, T. Hianik, G.-P. Nikoleli, Stabilized lipid films in electrochemical biosensors, Electroanalysis 22 (2010) 2747–2763.
[47] R. George, R.N.A.H. Lewis, R.N. McElhaney, Reconstitution of the purified (Na^{+} + Mg^{2+})-ATPase from *Acholeplasma laidlawii* B membranes into lipid vesicles and a characterization of the resulting proteoliposomes, Biochim. Biophys. Acta 903 (1987) 283–291.
[48] T. Hianik, P. Rybar, G.M. Kostner, A. Hermetter, Molecular acoustic as a new tool for the study of biophysical properties of lipoproteins, Biophys. Chem. 67 (1997) 221–228.
[49] T. Hianik, S. Küpcü, U.B. Sleytr, P. Rybar, R. Krivanek, U. Kaatze, Interaction of crystalline bacterial cell surface proteins with lipid bilayers in vesicles, Colloids Surf. A 147 (1999) 331–339.
[50] A.P. Sarvazyan, Ultrasonic velocimetry of biological compounds, Annu. Rev. Biophys. Biophys. Chem. 20 (1991) 321–342.
[51] A.P. Sarvazyan, Development of methods of precise measurements in small volumes of liquids, Ultrasonics 20 (1982) 151–154.
[52] A.P. Sarvazyan, I.V. Chalikian, Theoretical analysis of an ultrasonic interferometer for precise measurements at high pressures, Ultrasonics 29 (1991) 119–124.
[53] O. Kratky, H. Leopold, H. Stabinger, The determination of the partial specific volume of proteins by the mechanical oscillator technique, in: E. Grell (Ed.), Methods in Enzymology, vol. 27. Academic Press, London, 1973, pp. 98–110.
[54] S.H. White, W.C. Wimley, Hydrophobic interactions of peptides with membrane interfaces, Biochim. Biophys. Acta 1376 (1998) 339–352.
[55] J.A. Killian, Hydrophobic mismatch between proteins and lipids in membranes, Biochim. Biophys. Acta 1376 (1998) 401–415.
[56] T. Hianik, U. Kaatze, D.F. Sargent, R. Krivanek, S. Halstenberg, W. Pieper, et al. A study of the interaction of some neuropeptides and their analogs with lipid bilayers, Bioelectrochem. Bioenerg. 42 (1997) 123–132.
[57] R. Krivanek, P. Rybar, E.J. Prenner, R.N. McElhaney, T. Hianik, Interaction of the antimicrobial peptide gramicidin S with dimyristoyl-phosphatidylcholine bilayer membranes: a densimetry and sound velocimetry study, Biochim. Biophys. Acta 1510 (2001) 452–463.
[58] J.M. Davis, D.M. Clare, R.S. Hodges, M. Bloom, Interaction of a synthetic amphiphilic polypeptide and lipids in a bilayer structure, Biochemistry 22 (1983) 5298–5305.
[59] P.H. Axelsen, B.K. Kaufman, R.N. McElhaney, R.N.A.H. Lewis, The infrared dichroism of transmembrane helical polypeptides, Biophys. J. 69 (1995) 2770–2781.
[60] J.C. Huschilt, B.M. Millman, J.H. Davis, Orientation of a-helical peptides in a lipid bilayer, Biochim. Biophys. Acta 979 (1989) 139–141.
[61] E.J. Bolen, P.W. Holloway, Quenching of tryptophan fluorescence by brominated phospholipid, Biochemistry 29 (1990) 9638–9643.
[62] M.R. Morrow, J.C. Huschilt, J.H. Davis, Simultaneous modeling of phase and calorimetric behavior in an amphiphilic peptide/phospholipid model membrane, Biochemistry 24 (1985) 5396–5406.
[63] Y.-P. Zhang, R.N.A.H. Lewis, R.S. Hodges, R.N. McElhaney, Interaction of a peptide model of a hydrophobic transmembrane α-helical segment of a membrane protein with phosphatidylcholine bilayers. Differential scanning calorimetric and FTIR spectroscopic studies, Biochemistry 31 (1992) 11579–11588.

[64] K.P. Pauls, A.L. MacKay, O. Soderman, M. Bloom, A.K. Taneja, R.S. Hodges, Dynamic properties of the backbone of an integral membrane polypeptide measured by ^{2}H-NMR, Eur. Biophys. J. 12 (1985) 1–11.
[65] W.K. Subczynski, R.N.A.H. Lewis, R.N. McElhaney, R.S. Hodges, J.S. Hyde, A. Kusumi, Molecular organization and dynamics of 1-palmitoyl-2-oleoyl-phophatidylcholine bilayers containing a transmembrane α-helical peptide, Biochemistry 37 (1998) 3156–3164.
[66] C. Paré, M. Lafleur, F. Liu, R.N.A.H. Lewis, R.N. McElhaney, Differential scanning calorimetry and ^{2}H nuclear magnetic resonance and Fourier transform infrared spectroscopy studies of the effects of transmembrane α-helical peptides on the organization of phosphatidylcholine bilayers, Biochim. Biophys. Acta 1511 (2001) 60–73.
[67] H.M. Bouvrais, P. Meleard, T. Pott, K.J. Jensen, J. Brask, J.H. Ipsen, Softening of POPC membranes by magainin, Biophys. Chem. 137 (2008) 7–12.
[68] H.W. Huang, Molecular mechanism of antimicrobial peptides: the origin of cooperativity, Biochim. Biophys. Acta 1758 (2006) 1292–1302.
[69] H. Khandelia, J.H. Ipsen, O.G. Mouritsen, The impact of peptides on lipid membranes, Biochim. Biophys. Acta 1778 (2008) 1528–1536.
[70] S.K. Kandasamy, R.G. Larson, Binding and insertion of alpha-helical anti-microbial peptides in POPC bilayers studied by molecular dynamics simulations, Chem. Phys. Lipids 132 (2004) 113–132.
[71] M.O. Jensen, O.G. Mouritsen, G.H. Peters, Simulations of a membrane-anchored peptide: structure, dynamics, and influence on bilayer properties, Biophys. J. 86 (2004) 3556–3575.
[72] K. Jacobson, O.G. Mouritsen, G.W. Anderson, Lipid rafts: at a crossroad between cell biology and physics, Nat. Cell Biol. 9 (2007) 7–14.
[73] A.C. Rowat, J. Brask, T. Sparrman, K.J. Jensen, G. Lindblom, J.H. Ipsen, Farnesylated peptides in model membranes: a biophysical investigation, Eur. Biophys. J. 33 (2004) 300–309.
[74] G.F. Gause, M.G. Brazhnikova, Gramicidin S and its use in the treatment of infected wounds, Nature 154 (1944) 703.
[75] N. Izumiya, T. Kato, H. Aoyaga, M. Waki, M. Kondo, Synthetic Aspects of Biologically Active Cyclic Peptides: Gramicidin S and Tyrocidines, Halsted Press, New York, 1979, pp. 1–107.
[76] L.H. Kondejewski, S.W. Farmer, D. Wishart, C.M. Kay, R.E.W. Hancock, R. S. Hodges, Gramicidin S is active against both gram-positive and gram-negative bacteria, Int. J. Pept. Protein Res. 47 (1996) 460–466.
[77] R.E.W. Hancock, P.G.W. Wong, Compounds which increase the permeability of the *Pseudomonas aeruginosa* outer membrane, Antimicrob. Agents Chemother. 26 (1984) 48–52.
[78] E.J. Prenner, R.N.A.H. Lewis, R.N. McElhaney, The interaction of the antimicrobial peptide gramicidin S with lipid bilayer model and biological membranes, Biochim. Biophys. Acta 1462 (1999) 201–221.
[79] E.J. Prenner, R.N.A.H. Lewis, L.H. Kondejewski, R.S. Hodges, R.N. McElhaney, Differential scanning calorimetric study of the effect of the antimicrobial peptide gramicidin S on the thermotropic phase behavior of phosphatidylcholine, phosphatidyl-ethanolamine and phosphatidylglycerol bilayer membranes, Biochim. Biophys. Acta 1417 (1999) 211–223.
[80] R.N.A.H. Lewis, E.J. Prenner, L.H. Kondejewski, R.S. Hodges, C.S. Flach, R. Mendelsohn, et al. Fourier transform infrared spectroscopic studies of the interaction of the antimicrobial peptide gramicidin S with lipid micelles and with lipid monolayer and bilayer membranes, Biochemistry 38 (1999) 15193–15203.

[81] E.J. Prenner, R.N.A.H. Lewis, M. Jelokhani-Niaraki, R.S. Hodges, R. N. McElhaney, Cholesterol attenuates the interaction of the antimicrobial peptide gramicidin S with phospholipid bilayer membranes, Biochim. Biophys. Acta 1510 (2001) 83–92.

[82] D.P. Kharakoz, A. Colloto, K. Lohner, P. Laggner, Fluid-gel interphase line tension and density fluctuations in dipalmitoylphosphatidylcholine multilamellar vesicles: an ultrasonic study, J. Phys. Chem. 97 (1993) 9844–9851.

[83] W. Schaaffs, Molekularakustik, Springer, Berlin, 1963.

[84] W.W. van Osdol, M.L. Johnson, Q. Ye, R.L. Biltonen, Relaxation dynamics of the gel to liquid crystalline transition of phosphatidylcholine bilayers: effects of chain length and vesicle size, Biophys. J. 59 (1991) 775–785.

[85] J.F. Nagle, D.A. Wilkinson, Lecithin bilayers: density measurements and molecular interactions, Biophys. J. 23 (1978) 159–175.

[86] E.J. Prenner, R.N.A.H. Lewis, K.C. Neuman, S.M. Gruner, L.H. Kondejewski, R.S. Hodges, et al. Nonlamellar phases induced by the interaction of gramicidin S with lipid bilayers, Biochemistry 36 (1997) 7906–7916.

[87] R. Zidovetzki, U. Banerjee, D.W. Harrington, S.I. Chan, NMR study of the interactions of polymyxin B, gramicidin S, and valinomycin with dimyristoyllecithin bilayers, Biochemistry 27 (1988) 5686–5692.

[88] H. Susi, J. Sampugna, J.W. Hampson, J.S. Ard, Laser-Raman investigation of phospholipid–polypeptide interactions in model membranes, Biochemistry 18 (1979) 297–301.

[89] T. Katsu, H. Kobayashi, Y. Fujita, Mode of action of gramicidin S on *Escherichia coli* membranes, Biochim. Biophys. Acta 860 (1986) 608–619.

[90] Y. Yagi, S. Kimura, Y. Imanishi, Interaction of gramicidin S analogs with lipid bilayer membranes, Int. J. Pept. Protein Res. 36 (1990) 18–25.

[91] D. Mihailescu, L.I. Horvath, Molecular dynamics of lipid association at the hydrophobic interface of gramicidin S, Eur. Biophys. J. 28 (1998) 216–221.

[92] U. Harzer, B. Bechinger, Alignment of lysine-anchored membrane peptides under conditions of hydrophobic mismatch: a CD 15N and 31P solid-state NMR spectroscopy investigation, Biochemistry 39 (2000) 13106–13114.

[93] Y.-P. Zhang, R.N.A.H. Lewis, R.S. Hodges, R.N. McElhaney, FTIR spectroscopic studies of the conformation and amide hydrogen exchange of a peptide model of the hydrophobic transmembrane α-helices of membrane proteins, Biochemistry 31 (1992) 11572–11578.

[94] M.M. Sperotto, O.G. Mouritsen, Dependence of lipid membrane phase transition temperature on the mismatch of protein and lipid hydrophobic thickness, Eur. Biophys. J. 16 (1988) 1–10.

[95] M.K. Jain, Introduction to Biological Membranes, Wiley, New York, 1988.

[96] T.V. Chalikian, A.P. Sarvazyan, K.J. Breslauer, Hydration and partial compressibility of biological compounds, Biophys. Chem. 51 (1994) 89–109.

[97] A.P. Sarvazyan, D.P. Kharakoz, Acoustics study of conformational state of proteins in water, in: G.M. Frank (Ed.), Molecular and Cell Biophysics, Nauka, Moscow, 1977, pp. 93–106.

[98] D.P. Kharakoz, Protein compressibility, dynamic, and pressure, Biophys. J. 79 (2000) 511–525.

[99] T. Hianik, S. Poniková, J. Bagelova, M. Antalik, Specific volume and compressibility of human serum albumin–polyanion complexes, Bioorg. Med. Chem. Lett. 16 (2006) 274–279.

[100] R.N.A.H. Lewis, F. Liu, R. Krivanek, P. Rybar, T. Hianik, C.R. Flach, et al. Studies of the minimum hydrophobicity of α-helical peptides required to maintain a stable

transmembrane association with phospholipid bilayer membranes, Biochemistry 46 (2007) 1042–1054.

[101] H.J. Apell, How do P-type ATPases transport ions? Bioelectrochemistry 63 (2004) 149–156.

[102] M.M. Marcus, H.-J. Apell, M. Roudna, R.A. Schwendener, H.-G. Weder, P. Läuger, (Na^+ + K^+)-ATPase in artificial lipid vesicles: influence of lipid structure on pumping rate, Biochim. Biophys. Acta 854 (1986) 270–278.

[103] M. Esmann, D. Marsh, Lipid–protein interactions with the Na,K-ATPase, Chem. Phys. Lipid 141 (2006) 94–104.

[104] F. Cornelius, N. Turner, H.R.Z. Christensen, Modulation of Na,K-ATPase by phospholipids and cholesterol. II. Steady-state and presteady-state kinetics, Biochemistry 42 (2003) 8541–8549.

[105] P.L. Jørgensen, Isolation of (Na^+ plus K^+)-ATPase, Methods Enzymol. 32 (1974) 277–290.

[106] H.-J. Apell, M.M. Marcus, B.M. Anner, H. Oetliker, P. Läuger, Optical study of active ion transport in lipid vesicles containing reconstituted Na,K-ATPase, J. Membr. Biol. 85 (1985) 49–63.

[107] S. Kresak, P. Rybar, A.M. Rubtsov, O.D. Lopina, H.-J. Apell, M. Roudna, et al., The compressibility of Ca-ATPase–lipid complexes in native sarcoplasmic reticulum in the presence of ATP (2011) (in preparation).

[108] T. Hianik, P. Rybár, R. Krivánek, M. Petríková, M. Roudna, H.J. Apell, Specific volume and compressibility of bilayer lipid membranes with incorporated Na,K-ATPase, Gen. Physiol. Biophys. 30 (2011) 145–153.

[109] H. Hebert, P. Purhonen, K. Thomsen, H. Vorum, A.B. Maunsbach, Renal Na,K-ATPase structure from cryo-electron microscopy of two-dimensional crystals, Ann. N. Y. Acad. Sci. 986 (2003) 9–16.

[110] D. Marsh, Handbook of Lipid Bilayers, CRC Press, Boca Raton, FL, 1990.

[111] P.L. Jorgensen, K.O. Hakansson, S.J.D. Karlish, Structure and mechanism of Na,K-ATPase: functional sites and their interactions, Annu. Rev. Physiol. 65 (2003) 817–849.

[112] D.P. Kharakoz, A.A. Shlyapnikova, Thermodynamics and kinetics of the early steps of solid-state nucleation in the fluid lipid bilayer, J. Phys. Chem. B 104 (2000) 10368–10378.

[113] K. Gekko, Y. Hasegawa, Compressibility–structure relationship of globular proteins, Biochemistry 253 (1986) 6563–6571.

[114] O.G. Mouritsen, A. Boothroyd, R. Harris, N. Jan, T. Lookman, L. MacDonald, et al. Computer simulation of the main gel–fluid phase transition of lipid bilayers, J. Chem. Phys. 79 (1983) 2027–2041.

[115] M.M. Sperotto, O.G. Mouritsen, Monte Carlo simulation studies of lipid order parameter profiles near integral membrane proteins, Biophys. J. 59 (1991) 261–270.

CHAPTER THREE

Cytoskeletal Reorganization of Red Blood Cell Shape: Curling of Free Edges and Malaria Merozoites

Doron Kabaso,[1,2,*] Roie Shlomovitz,[1] Thorsten Auth,[3] Virgilio L. Lew,[4] *and* Nir S. Gov[1]

Contents

Abstract

Human red blood cells (RBCs) lack the actin–myosin–microtubule cytoskeleton that is responsible for shape changes in other cells. Nevertheless, they can display highly dynamic local deformations in response to external perturbations,

* Corresponding author: Tel.: +386 1 476 88 25; Fax: +386 1 476 88 50.
E-mail address: doron.kabaso@fe.uni-lj.si

[1] Department of Chemical Physics, Weizmann Institute of Science, Rehovot, Israel
[2] Laboratory of Biophysics, University of Ljubljana, Ljubljana, Slovenia
[3] Institute for Solid State Research, Research Centre Jülich, Jülich, Germany
[4] Department of Physiology, Development and Neuroscience, University of Cambridge, Cambridge, United Kingdom

Advances in Planar Lipid Bilayers and Liposomes, Volume 13
ISSN 1554-4516, DOI: 10.1016/B978-0-12-387721-5.00003-1

such as those that occur during the process of apical alignment preceding merozoite invasion in malaria. Moreover, after lysis in divalent cation-free media, the isolated membranes of ruptured ghosts show spontaneous inside-out curling motions at the free edges of the lytic hole, leading to inside-out vesiculation. The molecular mechanisms that drive these rapid shape changes are unknown. Here, we propose a molecular model in which the spectrin filaments of the RBC cortical cytoskeleton control the sign and dynamics of membrane curvature depending on two types of spectrin filaments. Type I spectrin filaments that are grafted at one end, or at both ends but not connected to the rest of the cytoskeleton, induce a concave spontaneous curvature. Type II spectrin filaments that are grafted at both ends to the cytoskeleton induce a local convex spontaneous curvature. Computer simulations of the model reveal that curling, as experimentally observed, can be obtained either by an overall excess of type I filaments throughout the cell, or by the flux of such filaments toward the curling edges. Divalent cations have been shown to arrest the curling process and Ca^{2+} ions have also been implicated in local membrane deformations during merozoite invasion. These effects can be replicated in our model by attributing the divalent cation effects to increased filament membrane binding. This process converts the curl-inducing loose filaments into fully bound filaments that arrest curling. The same basic mechanism can be shown to account for Ca^{2+}-induced local and dynamic membrane deformations in intact RBCs. The implications of these results in terms of RBC membrane dynamics under physiological, pathological, and experimental conditions are discussed.

1. Curling of Red Blood Cell Shape

The cortical cytoskeleton of red blood cells (RBCs) is a two-dimensional network of spectrin filaments attached to the internal domain of integral membrane proteins through a variety of intermediate protein links [1]. Under physiological conditions, the spectrin network plays a crucial role in maintaining the normal biconcave shape of the RBC, which allows RBCs to deform and pass through narrow capillaries in the microcirculation. Spectrin network is also essential to stabilize the spiculated RBC shape [2–5]. A dramatic example of cytoskeleton-driven local dynamic membrane shape changes is that documented during the process of spontaneous inside-out membrane curling and eventual vesiculation following cell lysis [6,7]. The unique feature of this process, which makes it an ideal system for studying the molecular mechanism that drives the local shape dynamics of the RBC membrane, is that it operates in a total absence of cytosolic components and external sources of energy, in distilled water media with no divalent cations, and only a trace of a proton buffer. After lysis is completed, the membrane at the free edge of the lytic hole is observed to curl outward in a process that takes several seconds at 37 °C [6]. Our basic assumption in this study is that the spontaneous curvature of the cell membrane is determined by the anchoring of the spectrin filaments, and therefore connects the molecular (tens of nanometers) and the cellular (several micrometers) length scales. Our model explains how the cortical cytoskeleton can control membrane conformations locally or globally. The model is applied to investigate the curling of isolated RBC membranes and the potential role of Ca^{2+} in arresting curling and mediating localized dynamic membrane deformations in intact RBCs.

2. Cytoskeleton-Induced Membrane Curvature

Curling and subsequent vesiculation of the membrane at the lytic hole have been suggested to correlate with the presence of the spectrin cytoskeleton at the inner membrane surface [6]. After vesiculation is complete, the spectrin actin cytoskeleton detaches and all spontaneous membrane motion ceases [8,9]. The curling phenomenon is not unique to RBCs; images of curled open membranes have been observed in membrane preparations from yeast and other mammalian cells under different experimental conditions [10], suggesting that cortical cytoskeleton components of eukaryotic cells can generate and sustain similar open membrane configurations. Under most experimental conditions, however, a lytic hole in any cell or in a protein-free synthetic lipid bilayer rapidly reseals [6,9,11,12]. In the case of RBCs, the free edge eventually fuses to form cylindrical and spherical multilayered vesicles [6]. Any source of spontaneous curvature in the membranes [13–18] will induce the observed curling, as we demonstrate in our simulations below. The experimental evidence in favor of this curvature being induced by the spectrin cytoskeleton is based on the following points: As long as the spontaneous curling of the free edges proceeds, there will be no release of spectrin–actin to the medium. Spectrin–actin is suddenly released only after vesiculation is completed and the vesicles display retention of sealing markers [9]. In addition, the spectrin filaments are adsorbed only on the inner side of the membrane and this asymmetry is maintained throughout the curling process (see Fig. 2B). Divalent cations arrest the curling process at any intermediate stage of the vesiculation process and also block spectrin–actin release [9]. Lysing of red cells in media with higher ionic strength and osmolality than used for spontaneous vesiculation (1–2 mM Na HEPES, pH 7.5, 0.1 mM EDTA) allows for slow spectrin–actin dissociation with minimal or no spontaneous vesiculation [1,8,20,21]. Under these conditions, the membrane has the appearance of a giant liposome with invisible opening, devoid of any spontaneous motion, and vesiculation of such spectrin-free liposomes can only be induced by the application of vigorous shearing forces [21].

Collectively, this evidence documents a strong association of spectrin–actin with curls and motion. Although it does not directly demonstrate that spectrin filaments drive the curling, it shows that the physical properties of the spectrin–actin-free membrane are not compatible with the detectable spontaneous curling motion.

The spectrin filaments of the cytoskeleton are mostly in the form of tetramers that form a triangular network, with the ends of the tetramers connected to the membrane through actin–band 4 complexes (the vertices of the network). Additionally, the tetramers are randomly anchored to the membrane at their midpoints by ankyrin–band 3 complexes. For simplicity, we neglect the ankyrin anchor complexes and treat the spectrin tetramers as single filaments that can be anchored with only one end attached (type I) or with both ends attached (type II; Fig. 1). Dynamic dissociation and rebinding have been proposed to explain membrane fluctuations and remodeling

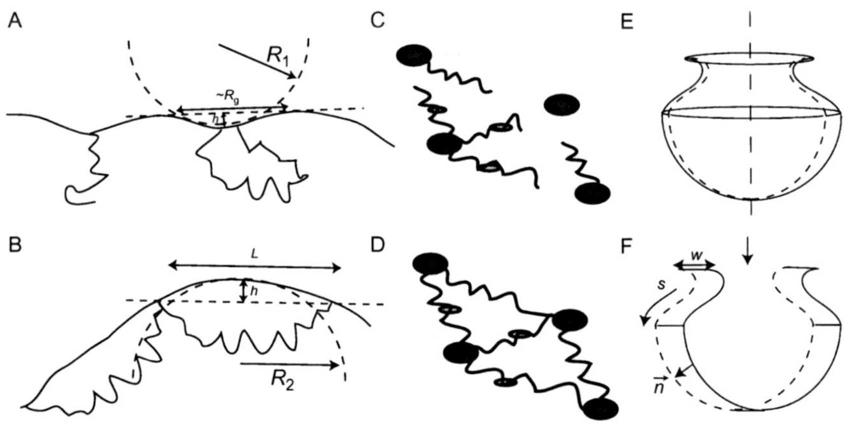

Figure 1 Schematic picture of the spontaneous curvature of the two types of spectrin filaments on the RBC membrane (adapted from Ref. [19]). (A) Type I spectrin filaments that are grafted at one end (left), or at both ends but not connected to the stretched network, induce a concave spontaneous curvature of radius R_1 (R_g is the radius of gyration of the filament). (B) Type II spectrin filaments that are grafted at both ends and are part of the connected stretched network that spans the RBC membranes inner surface (L is the separation between anchoring complexes) induce a local convex spontaneous curvature of radius R_2. An illustration of type I (C) and type II (D) filaments as part of a triangular network in which each edge represents a spectrin tetramer that can be connected to the membrane through actin–band 4 complexes (black circles) and can be further connected to the membrane through band 3 complexes (gray circles) is shown. In (C), the spectrin tetramers are anchored to membrane complexes only at one end (type I), whereas in (D) the spectrin tetramers are anchored at both ends and are part of the stretched network (type II). (E) Schematic illustration of the curling RBC ghost, which has a cylindrical symmetry, and the thin-strip approximation (F) that is used in our calculation. The strip has width w and a contour defined by the coordinate s, and the forces act everywhere along the local normal $\vec{n}$.

of the membrane after the solid-to-fluid transition of a cytoskeletal network [22,23]. Types I and II may be considered symbolic representations for alternative configurations of cytoskeletal components whose interplay can control dynamic changes in membrane curvature. This is the essential mechanistic content of our model. It has been shown both experimentally [24–27] and theoretically [28–30] that a polymer anchored with one end to a lipid bilayer, corresponding to type I spectrin filaments, exerts an entropic pressure such that the membrane curves away from the filament (defined here as a concave local shape with the normal pointing outward from the cell). The thermal fluctuations of the filaments are the source of this pressure and their configuration space is confined by the bounding membrane. This configuration space, and therefore the entropy, increases if the membrane bends away from the type I spectrin filament, thereby allowing more degrees of freedom for the filaments to explore different. The filament pulls the membrane at its anchor and pushes everywhere else. For a membrane patch of size πR_g^2, where the radius of gyration of the spectrin tetramer is $R_g = 13$ nm [31] and a membrane bending rigidity $\kappa = 10k_BT$, we estimate the induced curvature radius [32] to be $R_1 = 0.1$–1 μm for type I filaments (depending on the spectrin density and anchoring topology; see Fig. 1A). Type II filaments within the intact cytoskeleton network act as stretched entropic springs that locally curve the membrane toward the filament (defined here as a convex shape), as shown in Fig. 1. The finite distance between the spectrin cytoskeleton and the lipid bilayer [33], and the stretched spectrin bonds [34], determines the curvature imposed by the entropic pressure of type II filaments. Since in a disrupted network the anchor distance is unknown, we estimate the entropic effect of the filaments on the membrane curvature to be similar for both types of filaments, but with opposite sign, that is, $R_2 = -R_1\ 0.1$ μm (Fig. 1B).

The curvature induced by the spectrin network has been calculated on the length scale of spectrin mesh size [31], as well as on larger scales [35], and several experiments have demonstrated that anchored polymers can indeed produce a spontaneous curvature of such magnitude [26,27]. Based on these results, we developed a model to explain the membrane dynamics observed in the processes of curling and apical alignment. The basic assumption is that the variation in the concentrations of both types of membrane filaments controls the local membrane curvature in space and time (Fig. 2). Although we explicitly describe membrane dynamics in terms of type I and type II spectrin filaments in our model, the model will also apply when curvature is induced by membrane proteins other than spectrin. Divalent ions, such as Ca^{2+}, have been shown experimentally [9,36–38] to induce a stronger binding of the spectrin filaments to the membrane through their anchoring complexes. Therefore, we could translate the effect of divalent cations in the model as inducing a larger equilibrium fraction of type II filaments over type I filaments.

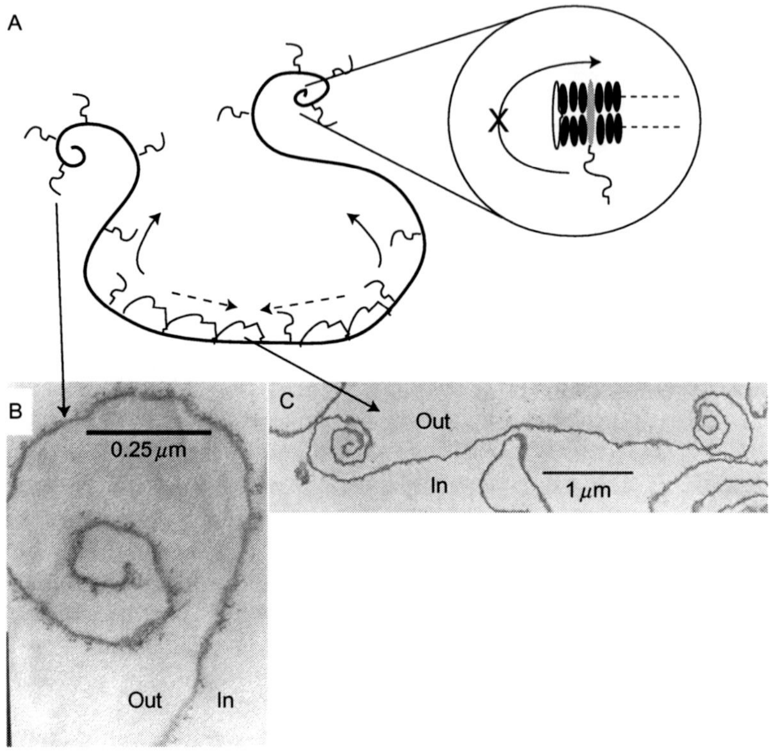

Figure 2 Schematic picture of our model (adapted from Ref. [19]). The connected network of filaments of convex spontaneous curvature (type II) is slowly shrinking and releasing its tension (dashed arrows), slowly squeezing out the type I filaments that have concave spontaneous curvature and migrate (solid arrows) to the curling lip at the open hole. Another mechanism is that rupture of the membrane during hole formation and the consequent curling process convert type II filaments to type I filaments, thereby increasing the overall excess density of type I filaments at the tips and throughout the ghost membrane. These mechanisms are proposed to explain the experimental patterns (A) [6]. In the circled inset is a representation of a type I filament at the hole rim, with lipids as black ovals and transmembrane anchor in gray. The pore edge can be stabilized by curved proteins that are denoted by an open gray shape. The asymmetry between the inner and outer faces of the membrane is maintained because filaments cannot diffuse through the rim to the outer side, and indeed filaments are seen only on the inner side of the membrane throughout the curling process (B). The time period of the curling process is within seconds of membrane rupture [6].

3. The Model for Curling of RBCs

In our model, the densities of both types of spectrin filaments control the membrane curvature (Figs. 1 and 2). We use a continuum model to compute the shape deformation of the membrane and the density distribution of the two types of filaments. The filament densities are treated as continuum variables, which are the averaged (coarse-grained) values of the discrete distribution of the real filaments. Although we explicitly use type I and type II spectrin filaments as the origin of the curvature, the model also applies when the curvature is induced by membrane proteins other than spectrin, such as curved integral proteins [13,39]. The bending effect of the filaments is treated here with a coarse-grained approach, ignoring the fine-scale wavy nature of the membrane shape due to the anchored filaments [28,31]. The average spontaneous curvature of the membrane is thus induced by the field of filaments. In a dilute regime, where excluded-volume interactions among the filaments are weak, this approach allows us to treat the induced curvature of the two types of filaments in a simple additive manner [40] in the equation for the membrane bending energy (Eq. 1). There are two possible ways to simplify the calculation (1) assume that the membrane deformation and filament distribution are cylindrically symmetric and (2) consider a thin strip of membrane (of width w) that is curved only along its length (while flat along the direction of the strip width—translational symmetry). We followed the second strategy. The Hamiltonian describing the free energy, E, is given by the standard Helfrich-type form [41]:

$$E = \frac{1}{2}\kappa w \int \left(\frac{\partial^2 r}{\partial s^2} - H_1 n_1 - H_2 n_2 \right)^2 \mathrm{d}s, \tag{1}$$

where s is the coordinate along the contour; n_1 and n_2 are the relative densities of the two types of filaments, normalized by the saturating (maximal) packing density of the filaments n_{sat}; κ is the bending rigidity modulus; and $H_1 = 1/R_1$ and $H_2 = 1/R_2$ are the two spontaneous curvatures associated with type I and type II filaments, respectively, so that the overall spontaneous curvature induced by the spectrin filaments is $n_1 H_1 + n_2 H_2$. This force arises from the mismatch between the local membrane curvature and the local value of the spontaneous curvature (due to the filaments), which drives the membrane to bend so as to reduce the mismatch. Using this free energy, we can now derive the equations of motion for the membrane shape and filament distribution.

Using the calculus of variation (see Supporting Material, Part S1), we derive the following force acting on the membrane in the normal direction, due to the instantaneous distributions of type I and type II filaments:

$$\begin{aligned} F_{\mathrm{n}} = \kappa \Bigg(& \left(\frac{\partial y}{\partial s}\frac{\partial^4 x}{\partial s^4} - \frac{\partial x}{\partial s}\frac{\partial^4 y}{\partial s^4} \right) + \frac{H_1}{2}\frac{\partial^2 n_1}{\partial s^2} + \frac{H_2}{2}\frac{\partial^2 n_2}{\partial s^2} \\ & + (n_1 H_1 + n_2 H_2)^2 \frac{H}{2} - \frac{3}{2} H^3 \Bigg), \end{aligned} \tag{2}$$

where the membrane curvature is $H = \partial x/\partial s(\partial^2 y/\partial s^2) - \partial y/\partial s(\partial^2 x/\partial s^2)$ (see Supporting Material for the derivation). This force arises from the mismatch between the local membrane curvature and the local value of the spontaneous curvature, which drives the membrane to bend so as to reduce the mismatch. The variation leading to Eq. (2) is a one-dimensional version of the more general expressions derived earlier [42,43]. These expressions recover the familiar form for small undulations of a flat membrane in the Monge gauge. Note that whereas most previous works dealt with the stationary shapes of membranes [42–44], here we are solving the dynamics of the membrane shape deformations. Below, we show that the dynamics is nontrivial; for example, a membrane with a uniform spontaneous curvature is shown to curl from its free edges inward. Details of the simulation techniques are given in the Supporting Material, and here we mention just the main properties. In addition to the curvature-driven forces given by Eq. (2), there are two more forces that act on the membrane in our simulations (1) the stiff harmonic-spring interaction between the nodes along the membrane maintains a constant overall length of membrane strip and (2) an artificial short-range repulsion prevents the membrane from overlapping on itself. The sum of these forces acting on the membrane is then used in the equation of motion of the membrane, which is written in the limit of local hydrodynamic interactions [45]:

$$V_{\mathrm{n}} = F_{\mathrm{n}}\xi, \tag{3}$$

where V_{n} is the normal velocity of the membrane and ξ is the effective friction coefficient that takes into account the viscous drag exerted by the fluid that surrounds the membrane locally (this is a simplified form of the Oseen tensor in the limit of only short-range hydrodynamic interactions). The full solution of this system, including the hydrodynamic flows around the curling membrane, awaits a future study. Finally, since we work in translational symmetry, the free edges of the membrane strip (the endpoints of our one-dimensional contour) correspond to the membrane edge along the hole in the real ghost. The hole radius and edge length are thus meaningless, and their energy contribution (line tension) is therefore constant. In the real situation, the hole starts from some minimal radius size, which has a large distribution. Nevertheless, the curling process is observed experimentally [6] to proceed without any correlation to the radius of the hole; it is observed to be identical along the edges of both small holes of radius $R \approx 0.5$ μm and large holes of $R \approx 10$ μm (Fig. 2C). We therefore decided to investigate the curling process independently of the issue of hole edge length, as is inherent to our choice of translational symmetry (Supporting Material, Part S2.4). The weak effective line tension observed in the RBC ghosts [6] may arise due to the stabilization of the pore edge by curved membrane proteins (Fig. 2A, inset). The simulations presented here using translational symmetry are reasonably accurate as long as the hole radius is much larger than the radius of curvature induced by the filaments, that is, $R \gg 0.1$ mm. The boundary conditions at the two free endpoints of the membrane (on both sides of the hole) were set, so that the membrane at the edges would have the curvature imposed by the local filament density (Supporting Material, Part S2.3). The filaments diffuse in the membrane, with conservation of the total number of filaments for each type. For simplicity, we do not allow spontaneous filament-type interconversion although this can easily be added. The membrane conformation acts as a potential landscape in which the two different types of filaments flow. The resulting fluxes of filaments are given in the Supporting Material.

4. Numerical Simulations

4.1. Effects of Filament Type and Density on Membrane Curling Pattern

To isolate the effect of filament types and densities on the membrane dynamics, we ran simulations with only one type of filament (i.e., type I or type II). The filaments in this set of simulations had negligible filament mobility ($n_{sat} = 1000\ \mu m^{-2}$; $\Lambda/n_{sat} = 1 \times 10^{-5}$ g s^{-1}, where Λ is the filament mobility in the membrane). Figure 3A and B shows snapshots of simulations with type II filaments at low and high relative densities. Curling at the free edges is inward, with an outside-in topology (a pattern that has been observed in a synthetic vesicle system [46] but never observed experimentally in RBCs), and the radius of the curled membrane is 0.25 μm and 0.1 μm for the initial relative filament densities of 0.4 and 1, respectively. Note that the curling process starts at the ends of the strip in our simulations, as observed experimentally [6]. This is because the curvature forces are mostly balanced within the bulk of the membrane, leaving the free edge as the only location of unbalanced forces. Mathematically, this can be seen from Eq. (2): For a uniform distribution of one type of filament, the force due to the filaments comes only from the fourth term on the right-hand side. This term is proportional to the membrane curvature, which is initially low everywhere (radius of the ghost membrane) except at the free ends, where the local curvature is defined by the spontaneous curvature of the filaments there (Part S2.3, Eq. S22 of the Supporting Material). The forces at the free ends are therefore dominant and drive the curling from the edges inward.

In Fig. 3C and D, we show the analogous results for type I filaments. Here, we find the same curvature radii of 0.25 and 0.1 μm, but curling is outward, with an inside-out topology. This agrees with the curling observed in the RBC and was also observed in the curling synthetic vesicle

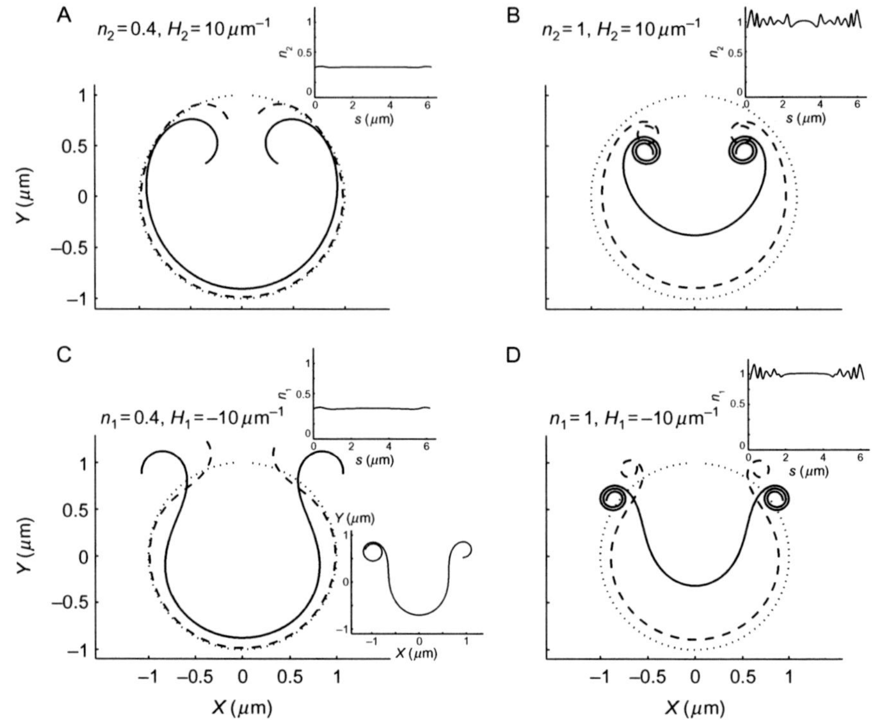

Figure 3 Effects of filament types and densities on the membrane dynamics (adapted from Ref. [19]). The two top panels (A) and (B) describe cases of type II filaments at low and high densities, within 30 s. The curvature sign is positive ($H_2 = 10\ \mu\mathrm{m}^{-1}$), and as a result the curling is inward. The two bottom panels (C) and (D) describe cases of type I filaments at low and high densities. The curvature sign is negative ($H_1 = -10\ \mu\mathrm{m}^{-1}$), and as a result the observed curling is outward. In all cases, the mobility of the filaments was quite low ($n_{\mathrm{sat}} = 1000\ \mu\mathrm{m}^{-2}$; $\Lambda/n_{\mathrm{sat}} = 1 \times 10^{-5}\ \mathrm{g\ s}^{-1}$). The insets at the top of each panel describe the final distribution of the filaments. Since the mobility is low, the final distribution has small fluctuations around the initial uniform spatial distribution. In each panel, the dotted line describes the initial membrane shape of a circle with a hole, the dashed line describes an intermediate time, and the solid line describes the final shape within a time period of 30 s.

system [46]. After 30 s, the membrane has curled three turns for simulations with high filament density, whereas not even a single turn was completed in the simulations at low density. If the simulation in Fig. 3C is allowed to run over a much longer time (90 min), we get curling with more than one turn of the membrane and an even greater radius of curvature ($\approx 0.2\ \mu\mathrm{m}$). This greater radius of curvature is due to the flux of filaments toward the edges (results not shown). The high-density case matches the experimentally observed radius of the innermost curl ($0.1\ \mu\mathrm{m}$; Fig. 3D). Note that excluded-volume repulsive forces generate the spiral pattern such that only the innermost part of the membrane achieves the desired curvature and the rest of the membrane spirals around it with an ever larger radius of curvature. These results show that only the simulations with a high density of type I filaments in the ruptured ghost render the experimentally observed curling pattern of the membrane (Fig. 3D). For a good experimental fit, $>90\%$ of the filaments should be of type I. This suggests that lytic rupture in the unique conditions that trigger spontaneous inside-out vesiculation explosively redistributes the filament distribution from a prelytic state with comparable proportions of type I and type II filaments to a $>90\%$ type I filament excess. Excesses of smaller magnitude would lengthen the curling process and reduce the curvature, and this may also account for some of the experimentally observed variability. Note that the membranes in the experiments evolved further beyond the curling process, by means of a series of cutting–splicing events, so that the curled part of the ghost is finally dashed up into small vesicles [6,7]. These vesicles satisfy the minimum curvature energy requirement as dictated by the density of type I filaments adsorbed on the external surface of the membrane. This step cannot be captured by our simplified calculation using only a thin strip of membrane.

4.2. Effects of the Initial Filament Distribution on the Membrane Curling Pattern

During rupture of the RBC, a higher proportion of type I filaments might be produced close to the lytic hole due to mechanical disruption of the spectrin network. To evaluate the effect of an inhomogeneous initial distribution of the filaments on the membrane dynamics, we ran simulations with asymmetric distributions for both types of filaments. In addition, the diffusion of spectrin filaments with high mobility ($n_{sat} = 100\ \mu m^{-2}$; $\Lambda/n_{sat} = 1 \times 10^{-4}\ g\ s^{-1}$) was included. The initial distributions had an excess of type I filaments at the free edges, with the excess decaying as a Gaussian from the edges to the middle. We varied the gradient of the filament density by changing the width of the Gaussian distribution. Far from the edges, we assumed that the two types of filaments have equal concentration, as in an intact RBC. The distribution of type I at the edge was set to a density of 0.7 or 0.83 (i.e., an excess amplitude of $\delta = 0.2$ or $\delta = 0.33$, respectively), decaying to a base level of 0.5 with a variance of either $\sigma = 1\ \mu m^2$ or $\sigma = 6\ \mu m^2$. The excess of type I filaments results from the detachment of type II filaments, so that an excess of one type occurs at the expense of the other type (conservation of total number of filaments). In Fig. 4, we show the results for different amplitudes and widths of the initial density variation. After 30 s of simulation, the results show that when the amplitude of the normal distribution is $\delta = 0.33$ filaments μm^{-1}, $\sigma = 1\ \mu m^2$ or $\delta = 0.33$ filaments μm^{-1}, $\sigma = 6\ \mu m^2$, the edges deform by half a turn and one complete turn, respectively. We observe that as the

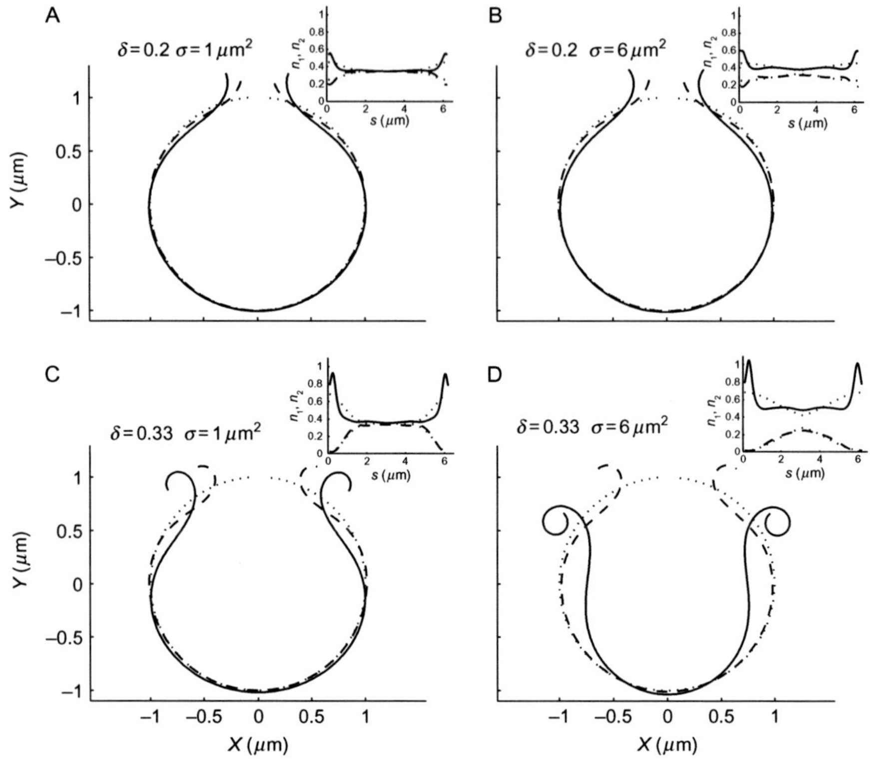

Figure 4 Effects of the spatial distribution of type I and type II filaments on membrane dynamics (adapted from Ref. [19]). The initial distribution has an excess of type I filaments localized at the edges of the hole, with a normal distribution of amplitude (δ) such that the density of the filaments at the edges is $0.35 \pm \delta$ (for type I and type II, respectively). The top and bottom rows are for $\delta = 0.2$ (μm^{-1}) and $\delta = 0.33$ (μm^{-1}), and the left and right columns are for $\sigma = 1$ (μm^2) and $\sigma = 6$ (μm^2), respectively. The mobility of the filaments is quite high ($n_{sat} = 100\ \mu m^{-2}$; $\Lambda/n_{sat} = 1 \times 10^{-4}$ g s^{-1}) in comparison to the low mobility in Fig. 3. Note that for the case of $\delta = 0.2$ (μm^{-1}), this amplitude was not sufficient to curl the membrane within the given time frame of 30 s (A, B). In addition, the case of high amplitude and low variance is not sufficient to form a complete curl. Finally, the case of high amplitude $\delta = 0.33$ (μm^{-1}) and high variance $\sigma = 6$ (μm^2) yields a single curl within 30 s. Note that the larger variance creates a larger pool from which to attract more type I filaments toward the edges (D). In each inset, the dotted lines are the initial distribution of filaments, the dashed line is the distribution of type II filaments, and the solid line is the distribution of type I filaments. In the inset of (D), note the large flux of filament type I toward the edges.

membrane edge curls, this curvature attracts type I filaments, whereas type II filaments are repelled from this region (see distributions of filaments in the insets). We conclude that if the mobility of the filaments is large enough to respond to the membrane curvature with a large flux, as shown in Fig. 4D, then a relatively moderate excess of type I over type II filaments at the edges of the hole is sufficient to curl the free edges of RBCs to the extent seen in the experiments. If the excess at the edges is too small (Fig. 4A–C), we find that the resultant membrane curling is much smaller compared to that in the experiments. We further conclude that diffusion and segregation of the two filament types are not sufficient by themselves to drive the curling of the whole membrane—a global excess of type I is also required.

4.3. Modeling the Effect of Divalent Ions

The documented Ca^{2+}-induced increase in the binding strength of spectrin filaments to the membrane [38] may be expected to effectively freeze the filament distribution due to a strong decrease in filament mobility, with an ensuing turnover of type I to type II filaments (Eq. S28) and cytoskeletal crosslinking [6]. We found that the curling and spontaneous inside-out vesiculation process could be instantly arrested at any stage by the addition of divalent cations to the medium, but reversibility was observed only with Mg^{2+} ions. In Fig. 5, we model these divalent cation effects under two of the conditions analyzed above (1) low mobility of spectrin filaments and large excess of type I filaments throughout the membrane (as in Fig. 3D) and (2) high-mobility filaments and a localized excess of type I filaments at the free edges (as in Fig. 4D). In the low-mobility case (top row), we simulated the effect of the divalent ions after allowing the membrane to curl freely for 2 s. We first modeled the effect of the ions by elimination of filament diffusion, but this did not stop the curling because the excess of type I filaments was still high enough to keep the process going (Fig. 5A). We next modeled the effect of the ions as a global turnover of type I to type II. We chose a significant turnover of 50% (this value is somewhat arbitrary) of type I to type II due to the addition of the ions and found that this practically arrested the curling process; the spontaneous curvature was reduced significantly (to zero for this choice of parameters) and the curls began to slowly unfold (Fig. 5B).

In the high-mobility and localized excess case (Fig. 5, bottom row), we tested the effects of the divalent ions by elimination of filament diffusion (Fig. 5C) and turnover of type I to type II (Fig. 5D). We found that when the ions diminished the diffusion, the curling process was considerably slowed down (Fig. 5C) compared to the results shown in Fig. 4D. A partial turnover of type I to type II also effectively arrested the curling (Fig. 5D), initiating a partial unfolding of the curls. We conclude that when the mobility of spectrin filaments is high enough to play a role in the curling process, the elimination of diffusion due to the ions will slow down the curling, whereas the turnover of type I to type II filaments due to the addition of ions is always a sufficient mechanism to account for the observed arrest in the curling.

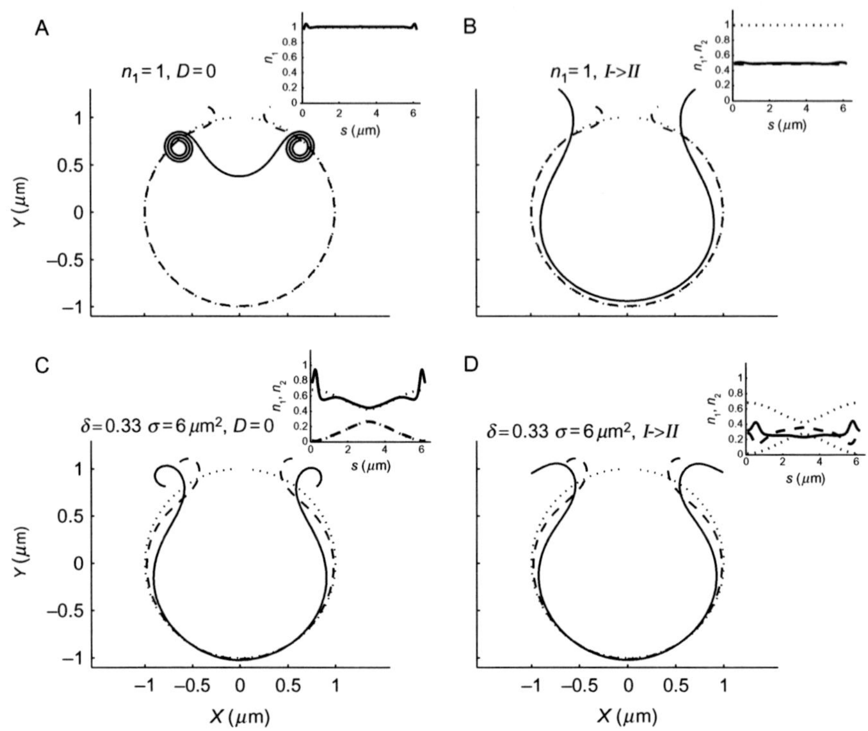

Figure 5 Modeling the effects of adding divalent ions on the membrane dynamics (adapted from Ref. [19]). The addition of divalent ions is modeled by eliminating the diffusion of filaments or by the turnover of type I to type II filaments. This modeling is performed for two successful cases (1) high and uniform density ($n_1 = 1$) of type I filaments with low mobility, as described in Fig. 3D (shown here in the top row) and (2) a more localized spatial distribution with large amplitude and variance ($\delta = 0.33$ (μm^{-1}), $\sigma = 6$ (μm^2)) as described in Fig. 4D (shown here in the bottom row). The results for the arrest of diffusion are demonstrated in the left column, and the results for the partial turnover of type I to type II filaments are demonstrated in the right column. Since the curling in Fig. 3D is obtained within 10 s, the arrest of diffusion is performed after 2 s (A). Note that the arrest of diffusion is not able to stop the curling. In contrast, the curling in Fig. 4D is obtained within minutes, which allows us to arrest the diffusion after 10 s (C). This arrest slows down the curling, and a single full curl is not obtained within the time frame of 30 s. In the right column, the turnover is of 50% of type I into type II filaments, which is sufficient to stop the curling and to even unfold its curled edges (B, D).

5. Malaria Merozoites and Ca^{2+} Dynamics

During the approach of a malaria merozoite to a targeted RBC, the red cell undergoes local and highly dynamic shape changes [47–50]. If the initial contact is away from the apical pole of the merozoite, which is required for penetration, the RBC responds with rapid merozoite-engulfing motions and the merozoite reorients to achieve apical contact. As soon as this is attained, the RBC returns to its normal biconcave shape and remains quiescent while it is penetrated by the parasite. Immediately after the parasite is internalized, a second and more prolonged wave of dynamic deformations takes place. Ca^{2+} ions have been implicated in both early and late dynamic responses during merozoite invasion [51].

6. The Model for RBC and Malaria Merozoite Interaction Prior to Invasion

The model as described up to now proved sufficient to describe the curling process of RBCs. In the case of the merozoite attachment, only a small patch of the RBC membrane close to the parasite has to be simulated, so we used periodic boundary conditions with increased spatial resolution compared to the simulations for curling. To simulate the effects attributed to Ca^{2+} in the malaria preinvasion events, we incorporated a simplified version of the known pump-leak Ca^{2+} balance across the RBC membrane.

The relation between the local Ca^{2+} concentration and the density of protein filament concentrations was assumed to be described by an instantaneous chemical equilibrium relation:

$$n_1 \rightleftharpoons n_2, \tag{4}$$

where the left and right arrows are for the equilibrium constant k and k', respectively. The two equilibrium constants are

$$\begin{aligned} k &= \exp\left(\frac{\alpha n_{\mathrm{Ca}}}{k_{\mathrm{B}}T}\right), \\ k' &= \exp\left(\frac{-\alpha n_{\mathrm{Ca}}}{k_{\mathrm{B}}T}\right), \end{aligned} \tag{5}$$

where αn_{Ca} represents the Boltzmann energy difference δE. The concentration of the two filament types is determined from

$$\begin{aligned} n_2 &= n_1 \exp\left(\frac{-2\alpha n_{\mathrm{Ca}}}{k_{\mathrm{B}}T}\right), \\ n_1 &= n_{\mathrm{tot}} - n_2, \end{aligned} \tag{6}$$

where the exponential factor is derived from k'/k (Eq. 5) and n_{tot} is the total density of filaments, which equals unity. The local density of type II filaments depends on the local density of Ca^{2+} as follows:

$$n_2 = n_{tot} \frac{1}{\exp(-0.05 n_{Ca}) + 1}, \tag{7}$$

where the numerical factor in the exponent gives the Ca^{2+}-induced affinity of spectrin filament binding. The density of Ca^{2+} ions is

$$n_{Ca} = N_{Ca}/\mathrm{d}s, \tag{8}$$

where N_{Ca} is the number of Ca^{2+} ions and ds is the distance between consecutive nodes.

The density of Ca^{2+} ions is treated along the membrane and represents those ions that reside within the interaction volume close to the membrane, where they can interact with the spectrin filaments. The thickness of this interaction volume is similar to that of the spectrin network, which is ~30 nm, and its width is simply that of the membrane strip that we are modeling. The influx of Ca^{2+} is assumed to occur only at the region of contact between merozoite coat and RBC membrane, which has a length of 0.5 μm. A single influx pulse of Ca^{2+} is modeled with maximum amplitude of ~600 ions μm^{-3}, to ensure that $n_2 \sim 1$ at the contact region, and is held constant for 2 s. After this time, the amplitude of the pulse decays exponentially by active extrusion and by diffusion away from the contact region along the membrane according to a simple diffusion equation:

$$\frac{\partial N_{Ca}}{\partial t} = D_{Ca} \frac{\partial^2 N_{Ca}}{\partial s^2} - K_{eff} N_{Ca} + J_{in}, \tag{9}$$

where D_{Ca} is the effective diffusion constant for Ca^{2+} ions, K_{eff} is the efflux rate of Ca^{2+} ions, and J_{in} is the influx of Ca^{2+}.

Most of the biophysical parameters used in this work were taken from the literature. The diffusion coefficient of proteins within the membrane was previously estimated to be 0.01–0.005 $\mu m^2\ s^{-1}$ [52], whereas the diffusion coefficient of spectrin filaments was smaller, at 0.001 $\mu m^2\ s^{-1}$ [52]. The bending rigidity is on the order of $10k_BT$. The only parameter that was fitted to give a good agreement with the experiment was the effective membrane friction in Eq. (3). The efflux of Ca^{2+} was taken as 35 s^{-1} [53] and the diffusion coefficient of Ca^{2+} was estimated at 1 $\mu m^2\ s^{-1}$ [54]. The saturating (maximal packing) density of the filaments, n_{sat}, can be estimated from previous measurements [38] and taken to be on the order 1000 μm^{-2}.

6.1. Modeling the Effects of Ca^{2+}-Dependent Filament Turnover

We next applied our model to simulate the membrane dynamics of an intact RBC in response to a transient local influx of Ca^{2+} ions. Such Ca^{2+} influx transients have been proposed to follow the initial contacts between merozoites and RBCs before merozoite invasion [48,51]. Ca^{2+} influx was simulated by means of a step-function amplitude with a spatial width of 0.5 μm held constant for 2 s. The density of the type I and type II filaments was determined by the local instantaneous Ca^{2+} concentration (see Eq. S28). We chose a 3-μm long membrane for these simulations, using periodic boundary conditions. We assumed that the initial filament density was the same for both species ($n_1 = n_2 = 0.5$). The Ca^{2+} pulse was assumed to cause a complete turnover of type I to type II filaments at the point of maximum amplitude, to give us an estimate of the upper bound of the amplitude of membrane deformation. The results are shown in Fig. 6A and B (Table 1). The Ca^{2+}-induced turnover of type I to type II filaments caused the bending of the initially planar membrane in both the inward and outward directions due to the resulting inhomogeneous spontaneous curvature. The maximal membrane deformation was 0.4 μm after 2 s. In Fig. 6C and D, the Ca^{2+} influx region was held fixed in space and prevented from folding inward. Despite this restriction, the folding outward of the surrounding membrane remained robust. Note that although the adhesion region prevents the membrane from folding inward, the folding outward remains robust (C).

7. Modeling the Egress of Malaria Merozoites

While previous research has shown that proteases induced by the parasite degrade the RBC proteins as well as the cytoskeleton [55–57], we here hypothesize that degradation of the spectrin network stimulates a turnover of type II into type I filaments (Fig. 7). In addition, type I filaments may stabilize membrane pores formed seconds before the lytic stage and the dispersion of packed merozoites. The rapid coalescence of smaller pores into larger pores may be facilitated by increasing densities of type I filaments (Fig. 7). Furthermore, the high densities of type I filaments following the lytic stage could explain the documented outward curling of the RBC membrane. Future modeling and experimental studies are needed to test these ideas.

8. Conclusions

The results presented here show that a model based on the assumption that spectrin filaments attached to the inner side of the lipid bilayer membrane induce a spontaneous curvature can explain local dynamic membrane

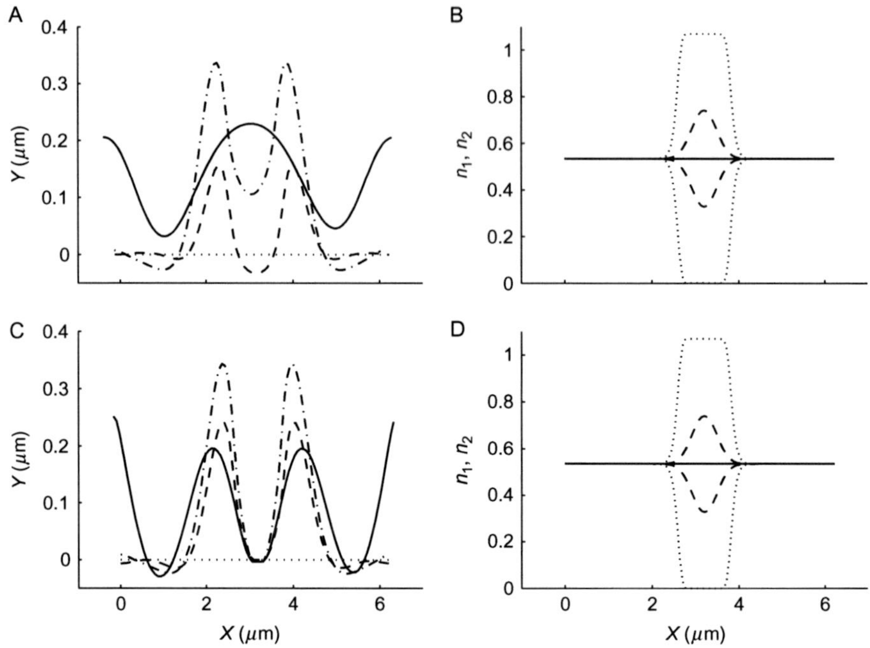

Figure 6 Exploring the effects of a Ca^{2+} influx pulse on local membrane shape in an intact RBC, assuming the same mechanism as that applied to explain curling arrest by divalent cations (adapted from Ref. [19]). Two scenarios are modeled (1) the membrane in the Ca^{2+} influx region is free to move (top row) and (2) it is held fixed (bottom row). The distributions of the two types of filaments are determined by the instantaneous local Ca^{2+} concentration. Therefore, the distribution has an excess of type II over type I filaments at the middle of the membrane. The density distributions of the two types are shown in (B, D). The influx of Ca^{2+} is constant for a period of 2 s and causes the conversion of type I to type II (dotted line in B, D), compared with an initial density of 0.5 for both types of filaments. After 2 s, the influx stops, and as a result the Ca^{2+} density decays within 0.5 s, and consequently the filament densities decay toward 0.5 (dashed line in B, D). The turnover of type I to type II filaments causes the bending of the straight initial membrane (dotted line in A, C) in both the inward and outward directions (D). After 0.5 s, the shape is given by the dashed line (A, C) and the maximal amplitude of 0.4 μm is reached after 2 s (the dotted-dashed line in A, C). The solid line is the shape of the membrane after 10 s. Slow mobility of the filaments is allowed and values are listed in Table 1.

shape changes observed in experiments. Two classes of filaments are considered in the model depending on whether one end (type I) or both ends (type II) of the spectrin tetramer are attached to the membrane. Each type induces a spontaneous curvature on the membrane of opposite sign and roughly equal magnitude. The model treats these nanometer-scale components in a continuum landscape of variable local densities of each filament type in response to environmental conditions and membrane shape. We sought

Table 1 List of parameters used in our simulation study

Parameter	Value	Parameter	Value
Curvature, H_1 (μm^{-1})	-10	Temperature, T (K)	300
Curvature, H_2 (μm^{-1})	10	Bending rigidity, κ ($k_B T$)	10
n_1, n_2, relative densities of type I and type II filaments	$\hat{n}_1/n_{sat}$ $\hat{n}_2/n_{sat}$	n_{sat}, maximum number of filaments per micron squared (μm^{-2})	100, 1000
Diffusion coefficient of filaments, D ($\mu m^2\ s^{-1}$)	0.001	Mobility of proteins, Λ ($g\ s^{-1}\ \mu m^{-2}$)	$D/(k_B T)$
Viscosity (water), η ($g\ m^{-1}\ s^{-1}$)	10^{-4}	Relative mobility ($g\ s^{-1}$)	Λ/n_{sat}
Filament length scale, d (μm)	0.1	Oseen tensor ($s\ g^{-1}$)	12.5
Threshold distance for repulsion (μm)	0.0327	D_{Ca} ($\mu m^2\ s^{-1}$)	1
Ca^{2+} efflux rate, K_{eff} (s^{-1})	35		

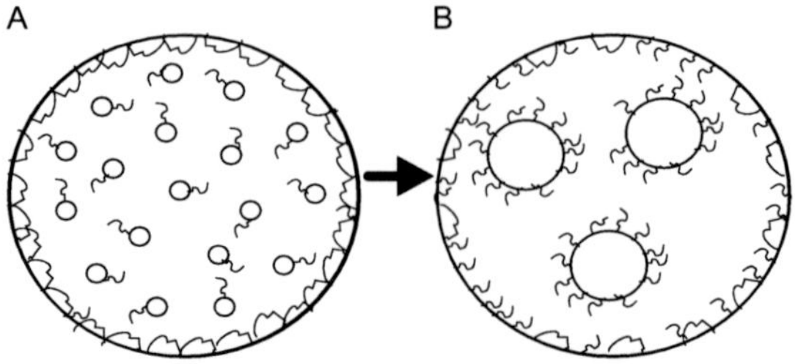

Figure 7 Schematic model for the stabilization and coalescence of membrane pores during egress of malaria parasites. (A) Small membrane pores (inner circles) are formed seconds before the egress and dispersion of malaria parasites. Since the pores are of a negative curvature, they can be stabilized by type I filaments, while type II filaments prefer the positive curvature of the red blood cell membrane (outer circle). (B) Large membrane pores are formed and stabilized by increasing densities of type I filaments. Note that the origin of type I filaments could be due to the degradation of type II filaments by proteases activated by the packed malaria parasites.

to determine whether the model could account for two well-documented examples of dynamic membrane deformations (1) the curling phenomenon of spontaneously vesiculating RBC ghosts and (2) the mechanism of membrane deformation during apical alignment of malaria parasites. In the study of the curling phenomenon, the simulations showed that either an overall excess of type I filaments or an initial excess of type I filaments close to the lytic hole could initiate the outward curling of the cell membrane. Comparing the model with experiment (see Figs. 3–5) allowed us to fit the free parameter of our model, the effective membrane hydrodynamic friction. Divalent cations have been shown to arrest the curling and spontaneous vesiculation process [6,9] and to induce stronger spectrin–membrane binding [37,38]. This effect is represented in the model by the elimination of filament diffusion and a turnover of type I to type II filaments. We find that the cation-induced turnover of type I to type II filaments is the dominant mechanism that allows us to reproduce the experimentally observed arrest of the curling (Fig. 5).

The mechanism of the effects attributed to Ca^{2+} was also investigated in relation to the potential of transient Ca^{2+} influx during merozoite apical alignment [47,48,51,58]. The results (Fig. 6) show that the amplitude and timescale of the membrane deformations induced by the transformation of type I to type II filaments are in good quantitative agreement with those documented in the best available observations of this process [47–49]. In addition, the simulations show that when the cytoskeleton consists mainly of type II filaments, membrane dynamics cannot occur and thus invasion efficiency is reduced. This may help explain the reduced invasion previously observed in ATP-depleted and volume-reduced RBCs [59–61]. The basic idea developed here—that spectrin filaments, or spectrin-like components of cortical cytoskeletons, control the local curvature of biological membranes and hence cell shape in static and dynamic conditions—is relevant not only to RBCs but also to all cellular systems with adsorbed networks of spectrin or spectrin-like filaments, such as neurons [50] and intracellular organelles [62].

Appendix Curling Ghosts

A1. Derivation of the Curvature Force on the Membrane

We derive the forces at the membrane, by treating it as a "one-dimensional membrane," that is, a thin strip of width w, with a bending modulus and tension coefficient. The Helfrich free energy of this membrane is given in Eq. (A7) and is used to derive the local restoring forces by the usual variation method. The derivation given below gives the details of this variation and is a one-dimensional version of the more general derivation given in Refs. [46,58].

Since the overall contour length is not constant in our system, the variation of the coordinates has to be taken with respect to their absolute index u along the contour, which is constant. In these terms, the curvature H appearing in the Helfrich free energy [59] is written as (standard differential geometry) x'

$$H = \frac{\dot{x}\ddot{y} - \ddot{x}\dot{y}}{\sqrt{\dot{x}^2 + \dot{y}^2}^3}, \tag{A1}$$

where the $\cdot$ symbol denotes differentiation with respect to the index of the point along the contour and the free energy is

$$\mathcal{F} = w\int^1 \left(\frac{1}{2}\kappa H^2 + \sigma\right)\sqrt{\dot{x}^2 + \dot{y}^2}\mathrm{d}u, \tag{A2}$$

where $\mathrm{d}s/\mathrm{d}u = \dot{s} = \sqrt{\dot{x}^2 + \dot{y}^2}$. The variation of this free energy gives the forces, for example, in the x-direction:

$$F_x = -\frac{\delta\mathcal{F}}{\delta x} = \frac{\mathrm{d}}{\mathrm{d}u}\frac{\partial\mathcal{F}}{\partial\dot{x}} - \frac{\mathrm{d}^2}{\mathrm{d}u^2}\frac{\partial\mathcal{F}}{\partial\ddot{x}}. \tag{A3}$$

The resulting equation of motion from this variation gives very long expressions that are not amenable to easy analysis, although they can be used for the numerical simulations.

To arrive at simpler expressions, we will develop the terms in Eq. (A3) and simplify at the end by assuming that the arc-length separation between the nodes along the contour are all the same. This is maintained as the simulation progresses by using the spline routine to rediscretize evenly the contour as its length evolves.

The first term on the RHS of Eq. (A3) is

$$\frac{\partial \mathcal{F}}{\partial \dot{x}} = H^2 \frac{\partial x}{\partial s} + 2H \frac{\partial H}{\partial \dot{x}} \dot{s}, \tag{A4}$$

where $\partial \dot{s}/\partial \dot{x} = \dot{x}/\sqrt{\dot{x}^2 + \dot{y}^2} = \partial x/\partial s$:

$$\begin{aligned} \frac{\mathrm{d}}{\mathrm{d}u}\frac{\partial \mathcal{F}}{\partial \dot{x}} &= 2\dot{s}HH'x' + \dot{s}H^2x'' + 2\dot{s}^2H'\frac{\partial H}{\partial \dot{x}} \\ &+2H\frac{\partial H}{\partial \dot{x}}\ddot{s} + 2H\dot{s}^2\frac{\partial H'}{\partial \dot{x}}, \end{aligned} \tag{A5}$$

where $\partial H/\partial \dot{x} = \ddot{y}/\dot{s}^3 - 3x'H/\dot{s}$. We therefore need to find an expression for $\ddot{y}$ (and $\ddot{x}$) by using the definition of $H = (x'\ddot{y} - y'\ddot{x})/\dot{s}^2$ and $\ddot{s} = x'\ddot{x} + y'\ddot{y}$. The final expressions that we get are $\ddot{y} = \ddot{s}y' + \dot{s}^2x'H$ and $\ddot{x} = \ddot{s}x' - \dot{s}^2y'H$.

We now assume that $\dot{s}$ is independent of u, so that $\ddot{s} = \dddot{s} = \cdots = 0$. The last term in Eq. (A5) becomes

$$\frac{\partial H'}{\partial \dot{x}} = -\frac{2}{\dot{s}}\frac{\partial}{\partial s}(x'H). \tag{A6}$$

The second term on the RHS of Eq. (A3) is

$$\frac{\partial \mathcal{F}}{\partial \ddot{x}} = \frac{-2Hy'}{\dot{s}} \tag{A7}$$

and we get

$$\frac{\mathrm{d}^2}{\mathrm{d}u^2}\frac{\partial \mathcal{F}}{\partial \ddot{x}} = -2\dot{s}\frac{\partial^2}{\partial s^2}(Hy'). \tag{A8}$$

We now write H in terms of s as $H = \vec{H}\cdot\hat{n} = -x''y' + y''x'$, where $\vec{H} = x''\hat{x} + y''\hat{y}$ and $\hat{n} = -y'\hat{x} + x'\hat{y}$, so that

$$H' = -y'x''' + x'y''',$$
$$H'' = -y''x''' + x''y''' - y'x^{(4)} + x'y^{(4)}. \tag{A9}$$

There is another force contribution from the membrane tension, giving a term of the form $F_x \propto \dot{s}x''$, so in the normal direction we get $F_n = \dot{s}\sigma(-y'x'' + x'y'') = \dot{s}\sigma H?$, where we used the identity $x'^2 + y'^2 = 1$ and therefore $2(x'x'' + y'y'') = \partial\ (x'^2 + y'^2)/\partial\ s = 0$.

Putting everything together, the normal force acting on the membrane due to curvature and tension is

$$F = \vec{F} \cdot \hat{n} = -y'F_x + x'F_y \tag{A10}$$

$$= \dot{s}\frac{1}{2}\kappa\left(-2\nabla^2 H - 2H\left(y'y^3 + x'x^3\right) - 3\left(\vec{H}\cdot\hat{n}\right)^3\right) + \dot{s}\sigma H \tag{A11}$$
$$= \dot{s}\kappa\left(-\nabla^2 H - \frac{1}{2}\left(\vec{H}\cdot\hat{n}\right)^3\right) + \dot{s}\sigma H$$
$$= \dot{s}\frac{1}{2}\kappa\left(2\left(y'x^{(4)} - x'y^{(4)}\right) - 3\left(\vec{H}\cdot\hat{n}\right)^3\right) + \dot{s}\sigma H,$$

where we used the identity $x'^2 + y'^2 = 1$ and therefore $2(x'x'' + y'y'') = \partial\ (x'^2 + y'^2)/\partial s = 0$ and $(y'y^3 + x'x^3) = -\ (x''^2 + y''^2) = -\ H^2$. These expressions are identical to those derived in Refs. [42,43].

Since the forces are per unit length, while we calculated above the forces per unit u, so we divide by $\dot{s}$ and finally get

$$F_n = \kappa\left(\left(y'x^{(4)} - x'y^{(4)}\right) - \frac{3}{2}H^3\right) + \sigma H. \tag{A12}$$

A1.1. Proteins with Spontaneous Curvature

When there are proteins with spontaneous curvature, the free energy (Eq. A2) changes to

$$\mathcal{F} = w\int^1 \left(\frac{1}{2}\kappa(H + \bar{H}n)^2 + (\sigma - \alpha n)\right)\sqrt{\dot{x}^2 + \dot{y}^2}\mathrm{d}u, \tag{A13}$$

where n is the density of proteins along the contour, which may not be uniform. Expanding the quadratic term, we get $H^2 + 2H\bar{H}n + (\bar{H}n)^2?$.

The variation of the first term was done above (all the variations are of the integrand times the $\dot{s}$ factor).

The new contributions to the forces acting on the membrane are (normal force per unit length)

$$F_{\text{spon,n}} = \overrightarrow{F_{\text{spon}}} \cdot \hat{n} = -y' F_{\text{spon},x} + x' F_{\text{spon},y} = \frac{\kappa}{2}\left((\bar{H}n)^2 H + 2(\bar{H}n'')\right) \tag{A14}$$

and

$$F_{\text{tension,n}} = \overrightarrow{F_{\text{spon}}} \cdot \hat{n} = -\alpha n H. \tag{A15}$$

A1.2. Fluxes and Diffusion of the Proteins

The conservation equation for the proteins, along the contour [42], becomes

$$\frac{1}{\dot{s}}\frac{\partial \dot{s}n}{\partial t} = \frac{D}{\dot{s}}\nabla_s^2(\dot{s}n) + \frac{1}{\dot{s}}\frac{\Lambda}{n_s}\nabla_s\left(\dot{s}n\nabla_s\left(\frac{1}{\dot{s}}\frac{\delta E}{\delta n}\right)\right), \tag{A16}$$

where the number of proteins in each unit contour length is $N = \dot{s}nn_s$ (n_s is the saturation concentration of the proteins), E is the energy functional of Eq. (A13), and the derivative along the contour is $\nabla_s = \nabla_u/\dot{s}$. We therefore get

$$\frac{\partial n}{\partial t} + \frac{n}{\dot{s}}\frac{\partial \dot{s}}{\partial t} = \frac{D}{\dot{s}}\nabla_s^2(\dot{s}n) + \frac{D}{k_B T n_s \dot{s}^2}\nabla_u\left(n\nabla_u\frac{\delta e}{\delta n}\right), \tag{A17}$$

where e is the energy per unit length, that is, the integrand in Eq. (A13) with respect to ds.

If the number of proteins is conserved, even though we allow the membrane overall length to change, then Eq. (A17) is correct. If however there is a reservoir of membrane that allows it to change in length, then this membrane can include lipids and proteins, so that the total number of proteins is not conserved when the membrane length changes. In this case, the change in the density due to length changes is removed, assumed to be balanced by the currents into/out of the reservoir. Equation (A17) is then modified by removing the second term on the left-hand side.

In our calculation of osteointegration, a nonlinear tension was employed, and as a result, the length of each membrane segment changed very little, so the second term on the LHS of Eq. (A17) was neglected.

A2. Numerical Realization of the Model

A2.1. Discretization of the Model

Since the flat shape model represents a segment of the whole cell, we used periodic boundary conditions. Thus, the number of grid points N equals the number of discretizations. In our model, the density n of element i is given by

$$n_i = \frac{N_i}{\Delta s_i}. \tag{A18}$$

A2.2. The Boundary Conditions

We employed periodic boundary conditions. The calculation of the first and second derivatives of the function along the x-direction was performed using the following explicit Euler method:

$$\begin{aligned} \frac{\partial x}{\partial s} &= \frac{x_{n+1} - x_{n-1}}{2\Delta s_n}, \\ \frac{\partial^2 x}{\partial s^2} &= \frac{x_{n-1} - 2x_n + x_{n+1}}{\Delta s_n^2}, \end{aligned} \tag{A19}$$

where the subscripts n, $n+1$, and $n-1$ represent the current, next, and previous nodes, respectively. The derivatives of the function along the y-direction were calculated in a similar manner. For the calculations of derivatives of the first point, the last point was added before it; while for the calculation of derivatives of the last point, the first point was added after it.

A2.3. The Variation of the Helfrich Equation

To find the slowest optimal projectile with the lowest energy of the membrane, we found the variation of the energy equation along the x and y coordinates, $\delta F/\delta x$ and $\delta F/\delta y$, respectively.

Our derivation as described above was verified using Mathematica software. The projected normal force along the x- and y-axes is

$$\begin{aligned} F_x &= F_{\mathrm{n}}\left(-\frac{\partial y}{\partial s}\right), \\ F_y &= F_{\mathrm{n}}\left(\frac{\partial x}{\partial s}\right), \end{aligned} \tag{A20}$$

where F_x and F_y are the projected forces along the x- and y-directions, respectively. The time evolution of the x and y coordinates is given by

$$x_i = x_i - \Delta t\left(1/2\kappa F_{x,i}\right)/\xi, \\ y_i = y_i - \Delta t\left(1/2\kappa F_{y,i}\right)/\xi, \tag{A21}$$

where $F_{x,i}$ and $F_{y,i}$ are contributions of the variation along the x and y coordinates, at index i, and ξ is the friction coefficient, which takes into account drag forces by the surrounding medium. The energy contributions $F_{x,i}$ and $F_{y,i}$ were taken only at the normal direction. Therefore, the membrane dynamics due to the energy minimization was constrained only in the normal direction.

A3. Analytic Derivation of the Steady-State Solution

We derive the steady-state solution in order to shed light on the underlying mechanisms responsible for the observed steady-state shape. In a steady state, the sum of fluxes of membrane-bound protein complexes (PCs) is equal to zero. In particular, the sum of the attraction flux due to membrane intrinsic curvature, J_{curv}, and the dispersion flux, J_{disp}, gives

$$\frac{\kappa\Lambda\bar{H}}{n_s}\left(\nabla H - \frac{\bar{H}}{n_s}\nabla n\right) = 0. \tag{A22}$$

The nontrivial solution is

$$\nabla n = \frac{n_s}{\bar{H}}\nabla H, \tag{A23}$$

where H is $\nabla^2 h$. By integrating both sides of Eq. A23 we get

$$n(x) = \frac{\nabla^2 h}{\bar{H}} + n_0. \tag{A24}$$

The function representing the peak region at the steady state is approximated up to a fourth-order polynomial. Note that due to symmetry around the peak as observed in numerical simulations, the odd orders of the polynomial are assumed to be zero. We use the proportionality found from Eq. A24 to obtain the shape function, $h(x)$, and the distribution of PCs, $n(x)$, as follows:

$$h(x) = h + fx^2 + gx^4, \\ n(x) = n_0 + 2f\frac{n_s}{\bar{H}} + \frac{n_s}{\bar{H}}gx^2, \tag{A25}$$

where x is the length along the x-axis. The following is the sum of forces F_{tot} derived from the free energy (Eq. (A13)):

$$F_{\text{tot}} = \kappa\left(-\nabla^2 H + \bar{H}\nabla^2 n + \frac{1}{2}n^2\bar{H}^2 H - \frac{1}{2}H^3\right) \\ +(\sigma - \alpha n)H - 2\gamma h. \quad \text{(A26)}$$

We find the steady-state analytic solutions for Eq. (A26). The aim is to have a better understanding of how changes in α impact on the SS shape and distribution of PCs. The steady states $n(x)$ and $h(x)$ (Eq. (A25)) were incorporated into the force equation (Eq. (A26)). After incorporating these two functions, we look for a solution that vanishes all the coefficients of zeroth-, second-, and fourth-order terms. The result of this system of equations is the solution for the three unknown functions: f, g, and h. The following is the list of parameter values used in this derivation: $\alpha = 0.05\ \text{g s}^{-2}$, $\gamma = 0.00004\ \text{g s}^{-2}$, $n_0 = 0.1$ (which is the initial relative density of PCs), $n_{\text{sat}} = 10\ \mu\text{m}^{-2}$, $\kappa = 100k_{\text{B}}T$, $\bar{H} = -10\,\mu\text{m}^{-1}$, and $\sigma = 0.001\ \text{g s}^{-2}$.

The first solution is the trivial solution with unknowns equal zero and the second solution is $h(x) > 0$, $g(x) < 0$, and $f(x) < 0$. The plot of this solution with respect to x gives a similar shape to the steady states seen in numerical simulations. From the approximation in (Eq. (A25)), the value of $1/\sqrt{f}$ can be derived to give the width of $h(x)$. We will next determine which parameter has the largest effect on the width of $h(x)$. From our analytic derivation, we obtain that

$$f_\alpha = \frac{-\bar{H}\left(2(\sigma - n_0\alpha) + \bar{H}^2 n_0^2 \kappa\right)}{16\left(\alpha - \bar{H}^2 n_0 \kappa\right)}, \quad \text{(A27)}$$

whereby neglecting the terms with κ due to their smaller magnitude yields:

$$f_\alpha \simeq \bar{H}\frac{n_0\alpha - \sigma}{8\alpha}. \quad \text{(A28)}$$

We find that the proportionality of the width is

$$1/\sqrt{f} = \sqrt{\frac{8\alpha}{\bar{H}(\sigma - n_0\alpha)}}. \quad \text{(A29)}$$

The following correlations are gleaned from Eq. (A29). Given a large adhesion constant (α) due to stronger adhesion to the extracellular matrix, the width of $h(x)$ is smaller to result in a sharper tip.

REFERENCES

[1] V. Bennett, The spectrin–actin junction of erythrocyte membrane skeletons, Biochim. Biophys. Acta 18 (1989) 107–121.

[2] A. Iglic, A possible mechanism determining the stability of spiculated red blood cells, J. Biomech. 30 (1997) 35–40.

[3] A. Iglic, V. Kralj-Iglic, H. Hagerstrand, Amphiphile induced echinocyte–spheroechinocyte red blood cell shape transformation, Eur. Biophys. J. 27 (1998) 335–339.

[4] A. Iglic, V. Kralj-Iglic, H. Hagerstrand, Stability of spiculated red blood cells induced by intercalation of amphiphiles in cell membrane, Med. Biol. Eng. Comput. 36 (1998) 251–255.

[5] H. Hagerstrand, M. Danieluk, M. Bobrowska-Hagerstrand, A. Iglic, A. Wrobel, B. Isomaa, M. Nikinmaa, Influence of band 3 protein absence and skeletal structures on amphiphile- and Ca^{2+}-induced shape alterations in erythrocytes: a study with lamprey (*Lampetra fluviatilis*), trout (*Oncorhynchus mykiss*) and human erythrocytes, Biochim. Biophys. Acta 1466 (2000) 125–138.

[6] V.L. Lew, A. Hockaday, R.M. Bookchin, Mechanism of spontaneous inside-out vesiculation of red cell membranes, J. Cell Biol. 106 (1988) 1893–1901.

[7] V.L. Lew, S. Muallem, C.A. Seymour, Properties of the Ca^{2+}-activated K channel in one-step inside-out vesicles from human red cell membranes, Nature 296 (1982) 742–744.

[8] T.L. Steck, R.S. Weinstein, D.F. Wallach, Inside-out red cell membrane vesicles: preparation and purification, Science 168 (1970) 255–257.

[9] T. Tiffert, A.R. Hockaday, V.L. Lew, Cytoskeletal protein changes and morphology during spontaneous inside-out vesiculation of human red cell membranes, J. Physiol. 487 (1995) 99.

[10] A. Duran, E. Cabib, B. Bowers, Chitin synthetase distribution on the yeast plasma membrane, Science 203 (1979) 363–365.

[11] B.E. Burke, D.M. Shotton, Erythrocyte membrane skeleton abnormalities in hereditary spherocytosis, Br. J. Haematol. 54 (1983) 173–187.

[12] J.M. Gluck, M. Wittlich, B.W. Koening, Integral membrane proteins in nanodiscs can be studied by solution NMR spectroscopy, J. Am. Chem. Soc. 131 (2009) 12060–12061.

[13] J. Zimmerberg, M.M. Kozlov, How proteins produce cellular membrane curvature, Nat. Rev. Mol. Cell Biol. 7 (2006) 9–19.

[14] A. Hagerstrand, L. Mrowczynska, U. Salzer, R. Prohaska, A.K. Michelsenn, V. Kralj-Iglic, et al. Curvature dependent lateral distribution of raft markers in the human erythrocyte membrane, Mol. Membr. Biol. 23 (2006) 277–288.

[15] A. Iglic, M. Lokar, B. Babnik, T. Slivnik, P. Veranic, H. Hgerstrand, V. Kralj-Iglic, Possible role of flexible red blood cell membrane nanodomains in the growth and stability of membrane nanotubes, Blood Cells Mol. Dis. 39 (2007) 14–23.

[16] A. Iglic, T. Slivnik, V. Kralj-Iglic, Elastic properties of biological membranes influenced by attached proteins, J. Biomech. 40 (2007) 2492–2500.

[17] J. Jorgacevski, M. Fosnaric, N. Vardjan, M. Stenovec, M. Potokar, M. Kreft, et al. Fusion pore stability of peptidergic vesicles, Mol. Member. Biol. 27 (2010) 65–80.

[18] Š. Perutkova, V. Kralj-Iglic, M. Frank, A. Iglic, Mechanical stability of membrane nanotubular protrusions influenced by attachment of flexible rod-like proteins, J. Biomech. 43 (2010) 1612–1617.

[19] D. Kabaso, R. Shlomovitz, T. Auth, V.L. Lew, N.S. Gov, Curling and local shape changes of red blood cell membranes driven by cytoskeletal reorganization, Biophys. J. 99 (2010) 88.

[20] D.M. Shotton, The proteins of the erythrocyte membrane, in: J.R. Harris (Ed.), Electron Microscopy of Proteins, vol. 4. Academic Press, London, 1983, pp. 205–330.
[21] T.L. Steck, J.A. Kant, Preparation of impermeable ghosts and inside-out vesicles from human erythrocyte membranes, Methods Enzymol. 31 (1974) 172–180.
[22] N.S. Gov, Active elastic network: cytoskeleton of the red blood cell, Phys. Rev. E Stat. Nonlin. Soft Matter Phys. 75 (2007) 011921.
[23] J. Li, G. Lykotrafitis, S. Suresh, Cytoskeletal dynamics of human erythrocyte, Proc. Natl. Acad. Sci. USA 104 (2007) 4937–4942.
[24] C. Frank, H. Frielinghaus, D. Richter, Hydrophilic alcohol ethoxylates as efficiency boosters for microemulsions, Langmuir 24 (2008) 6036–6043.
[25] V. Nikolov, R. Lipowsky, R. Dimova, Behavior of giant vesicles with anchored DNA molecules, Biophys. J. 92 (2007) 4356–4368.
[26] I. Tsafrir, Y. Caspi, J. Stavans, Budding and tubulation in highly oblate vesicles by anchored amphiphilic molecules, Phys. Rev. Lett. 91 (2003) 138102.
[27] I. Tsafrir, D. Sagi, T. Arzi, J. Stavans, Pearling instabilities of membrane tubes with anchored polymers, Phys. Rev. Lett. 86 (2001) 1138–1141.
[28] T. Auth, G. Gompper, Self-avoiding linear and star polymers anchored to membranes, Phys. Rev. E Stat. Nonlin. Soft Matter Phys. 68 (2003) 051801.
[29] E. Eisenriegler, A. Hanke, S. Dietrich, Polymers interacting with spherical and rodlike particles, Phys. Rev. E 54 (1996) 1134–1152.
[30] R. Lipowsky, Flexible membranes with anchored polymers, Colloids Surf. A Physicochem. Eng. Asp. 128 (1997) 255–264.
[31] T. Auth, N.S. Gov, S.A. Safran, Filament networks attached to membranes: cytoskeletal pressure and local bilayer deformation, N. J. Phys. 9 (2007) 430.
[32] T. Auth, G. Gompper, Self-avoiding linear and star polymers anchored to membranes, Phys. Rev. E 68 (2003) 051801.
[33] T. Auth, S.A. Safran, N.S. Gov, Fluctuations of coupled fluid and solid membranes with application to red blood cells, Phys. Rev. E Stat. Nonlin. Soft Matter Phys. 76 (2007) 051910.
[34] N.S. Gov, S.A. Safran, Red blood cell membrane fluctuations and shape controlled by ATP-induced cytoskeletal defects, Biophys. J. 88 (2005) 1859–1874.
[35] R. Podgornik, Surface polymer network model and effective membrane curvature elasticity, Phys. Rev. E Stat. Phys. Plasmas Fluids Relat. Interdiscip. Topics 51 (1995) 3368–3375.
[36] S.C. Liu, L.H. Derick, J. Palek, Visualization of the hexagonal lattice in the erythrocyte membrane skeleton, J. Cell Biol. 104 (1987) 527–536.
[37] Y. Takakuwa, N. Mohandas, Modulation of erythrocyte membrane material properties by Ca^{2+} and calmodulin. Implications for their role in regulation of skeletal protein interactions, J. Clin. Invest. 82 (1988) 394–400.
[38] F. Liu, H. Mizukami, A. Ostafin, Calcium-dependent human erythrocyte cytoskeleton stability analysis through atomic force microscopy, J. Struct. Biol. 150 (2005) 200–210.
[39] T. Auth, G. Gompper, Budding and vesiculation induced by conical membrane inclusions, Phys. Rev. E Stat. Nonlin. Soft Matter Phys. 80 (2009) 031901–031910.
[40] M. Breidenich, R.R. Netz, R. Lipowsky, The shape of polymer-decorated membranes, Europhys. Lett. 49 (2000) 431–437.
[41] W. Helfrich, Elastic properties of lipid bilayers: theory and possible experiments, Z. Naturforsch. C 28 (1973) 693–703.
[42] R. Lipowsky, The conformation of membranes, Nature 349 (1991) 475–481.
[43] Z.C. Tu, Z.C. Ou-Yang, Lipid membranes with free edges, Phys. Rev. E Stat. Nonlin. Soft Matter Phys. 68 (2003) 061915.
[44] U. Seifert, K. Berndl, R. Lipowsky, Shape transformations of vesicles: phase diagram for spontaneous-curvature and bilayer coupling models, Phys. Rev. A 44 (1991) 1182–1202.

[45] A. Veksler, N.S. Gov, Phase transitions of the coupled membrane–cytoskeleton modify cellular shape, Biophys. J. 93 (2007) 3798–3810.
[46] E. Mabrouk, D. Cuvelier, M.H. Li, Bursting of sensitive polymersomes induced by curling, Proc. Natl. Acad. Sci. USA 106 (2009) 7294–7298.
[47] J.A. Dvorak, L.H. Miller, T. Shiroishi, Invasion of erythrocytes by malaria merozoites, Science 187 (1975) 748–750.
[48] P.R. Gilson, B.S. Crabb, Morphology and kinetics of the three distinct phases of red blood cell invasion by *Plasmodium falciparum* merozoites, Int. J. Parasitol. 39 (2009) 91–96.
[49] S. Glushakova, D. Yin, J. Zimmerberg, Membrane transformation during malaria parasite release from human red blood cells, Curr. Biol. 15 (2005) 1645–1650.
[50] L. Lencesova, A. O'Neill, M.P. Blaustein, Plasma membrane–cytoskeleton–endoplasmic reticulum complexes in neurons and astrocytes, J. Biol. Chem. 23 (2004) 2885–2893.
[51] V.L. Lew, T. Tiffert, Is invasion efficiency in malaria controlled by pre-invasion events? Trends Parasitol. 23 (2007) 481–484.
[52] M.H. Grati, M.E. Schneider, B. Kachar, Rapid turnover of stereocilia membrane proteins: evidence from the trafficking and mobility of plasma membrane Ca^{2+}-ATPase 2, J. Neurosci. 26 (2006) 6386–6395.
[53] V.L. Lew, N. Daw, T. Tiffert, Distribution of plasma membrane Ca^{2+} pump activity in normal human red blood cells, Blood 102 (2003) 4206–4213.
[54] J.J. Blum, G. Lawler, I. Shin, Effect of cytoskeletal geometry on intracellular diffusion, Biophys. J. 56 (1989) 995–1005.
[55] C. Gelhaus, R. Vicik, T. Schirmeister, M. Leippe, Blocking effect of a biotinylated protease inhibitor on the egress of *Plasmodium falciparum* merozoites from infected red blood cells, Biol. Chem. 386 (2005) 499–502.
[56] S. Glushakova, J. Mazar, M.F. Hohmann-Marriott, E. Hama, J. Zimmerberg, Irreversible effect of cysteine protease inhibitors on the release of malaria parasites from infected erythrocytes, Cell. Microbiol. 11 (2009) 95–105.
[57] R. Chandramohanadas, P.H. Davis, D.P. Beiting, M.B. Harbut, C. Darling, G. Velmourougane, et al. Apicomplexan parasites co-opt host calpains to facilitate their escape from infected cells, Science 324 (2009) 794–797.
[58] M. Treeck, S. Zacherl, T.W. Gilberger, Functional analysis of the leading malaria vaccine candidate AMA-1 reveals an essential role for the cytoplasmic domain in the invasion process, PLoS Pathog. 5 (2009) e1000322.
[59] A.R. Dluzewski, D. Zicha, W.B. Gratzer, Origins of the parasitophorous vacuole membrane of the malaria parasite: surface area of the parasitized red cell, Eur. J. Cell Biol. 68 (1995) 446–449.
[60] J. Olson, A. Kilejian, Involvement of spectrin and ATP in infection of resealed erythrocyte ghosts by the human malaria parasite, *Plasmodium falciparum*, J. Cell Biol. 95 (1982) 757–762.
[61] T. Tiffert, V.L. Lew, N. Mohandas, The hydration state of human red blood cells and their susceptibility to invasion by *Plasmodium falciparum*, Blood 105 (2005) 4853–4860.
[62] M. De Matteis, J. Morrow, Spectrin tethers and mesh in the biosynthetic pathway, J. Cell Sci. 113 (2003) 2331–2343.

CHAPTER FOUR

Photovoltaic Solar Energy Conversion in Biomembranes: General Principles and Model System Studies☆

Felix T. Hong*

Contents

* Corresponding author: Tel.: +001 (313) 577-1538; Fax: +001 (313) 577-5494.
E-mail address: fhong@med.wayne.edu

Department of Physiology, Wayne State University, Detroit, Michigan, USA

☆ Dedicated to the memory of the late Prof. H. Ti Tien.

Advances in Planar Lipid Bilayers and Liposomes, Volume 13
ISSN 1554-4516, DOI: 10.1016/B978-0-12-387721-5.00004-3

Abstract

This chapter analyzes the general principles of photovoltaic solar energy conversion in biomembranes. Biological systems are inherently complex. However, unlike naturally occurring photosynthetic membranes, BLM-based photoconversion model systems are sufficiently simple to allow investigators to analyze the underlying photochemical and photophysical processes with mathematical rigor previously attainable only in physics and their progenies. We used a BLM-based model system of artificial electron pump to develop useful concepts and to devise a null current method, which can perform meaningful signal measurements that could be analyzed in terms of conventional concepts in electricity (equivalent circuit analysis). This method was then extended to the analysis of model membranes reconstituted from bacteriorhodopsin, which is perhaps the simplest photosynthetic pigment. Comparison of photosynthetic membranes with their non-biological counterparts yielded some interesting insights. Both the silicon photodiode and photosynthetic membranes start the conversion process with light-driven charge separation in an anisotropic environment. Whereas rectification is a key feature of the silicon photodiode for minimizing unwanted internal charge recombination, biological systems apparently used a different strategy to accomplish the same purpose. Illumination opens a proton conduction channel in bacteriorhodopsin, and cessation of illumination shuts off the conduction channel completely. Indirect evidence indicates that chlorophyll-based photosynthetic membranes also adopt the same strategy to prevent internal charge recombination in the dark, thus greatly enhancing the efficiency and effectiveness of photoconversion.

1. Introduction

Among problems facing the world in the twenty-first century, the issue of sustainable energy sources ranks high in the agenda. Moreover, the issue of energy sources is a major primary cause for problems facing humans: wars and deteriorating environments. The current major energy source—petroleum—is not distributed evenly among various geographic regions in accordance with population densities. Competition to the access of petroleum often threatens peace. The quest for alternative means for its access led to deep sea drilling, thus greatly increasing the risk of environmental pollution. Since petroleum is a nonrenewable energy source, all energy-related issues are going to get worse unless viable alternative sources are found soon. The rapidly increasing carbon dioxide level generated by consumption of oil and coal accentuates the acute need to find clean and renewable source of energy.

To make a long argument short, the only conceivable solution is to tap into solar energy [1,2]. It is clean and renewable as long as the Sun shines. It is more evenly and equitably distributed among the world; access to it threatens no others. Solar energy is the ultimate source of energy that is the direct precursor of hydraulic and wind energy. It was also the indirect precursor of oil and coal generated eons ago.

The present obstacle to the utilization of solar energy is mainly the cost or, more precisely, finding the right technology to harvest solar energy that is sufficiently economic to be competitive with oil and coal. A trivial but unacceptable solution is waiting—waiting till the price of oil and coal rises above that of solar energy production. Research and development of solar energy is not merely a science and technology problem. A sound long-term energy policy is sorely needed to foster timely research and development of alternative energy sources. Such an energy policy is however incompatible with the prevailing mentality of quarterly profits and instant gratification. Unlike the effort of putting humans on the Moon in the previous century, the science of solar energy conversion is not ready for immediate deployment for public utilization. How to make a long-range investment requires a great deal of political will of governmental officials and the public, especially in democratic countries. Take the recent predicament of the U.S. automobile industry for example. In good times, there was no incentive to explore alternative energy sources for automobiles because selling gas guzzlers was much more profitable than selling energy-efficient cars. In bad times, there was no incentive to do it because development of energy-efficient or gasoline-independent automobiles added costs to the production, thus deepening the economic crisis.

Ultimately, people must be educated to think beyond the simple-minded logic driven by the obsession of short-term benefits and instant gratification. The cost issue should not be an excuse for opposing the government-subsidized research and development of solar energy because almost all technological advances started as a costly, inconvenient, and often unreliable adventure, if not just a pipe dream. Without a long-range plan, we would never overcome the cost obstacle, not to mention the horror of getting caught at inconvenient time again and again. Clearly, it takes more than scientists and engineers together to overcome the obstacle. However, in spite of this pessimism, any promising news from the R and D sector helps raise the specter of a dream on the verge of realization, thus making well-meaning politicians' task easier and more realistic. With this brief digression, we can now address the science and technology aspect of biological solar energy conversion. The expertise of this author was limited to biological solar energy conversion and, in particular, photovoltaic solar energy conversion. The following discussion does not pretend to assert that the biological approach is the best approach. After all, we are still in the infancy of explorations. We merely pitch in our share of explorations.

In this chapter, we wish to examine the problem from the point of view of reverse-engineering Nature, that is, learning from Nature. Evolution was a long learning process that transcended the lifetime of individuals or even the existence of the human species. There was no guarantee that Nature would give us the best system. But there is reason to believe that learning from Nature offers an attractive shortcut in arriving at workable systems. In other words, imitating Nature constitutes a heuristic search of workable solutions, thus shortening our learning process. Whereas there were plenty of examples of showing how technology had been inspired by Nature, the process was not always a "one-way street." Sometimes, engineering ideas could enhance our understanding of biology. Insights into engineering aspects of man-made devices allow us to formulate the appropriate concepts and ask the right questions. This point is particularly relevant in the approach of model system studies. This discussion will not shy away from exploiting these two-way streets. We shall focus on the problem of photovoltaic solar energy conversion. Generation of hydrogen gas by splitting water via solar power is an attractive alternative to photovoltaic energy conversion [3,4], but the latter problem is beyond the scope of this chapter.

A model system is essentially a reduced natural system designed to ease the analysis of the corresponding natural system. Engineering concepts often help us decide what features to keep and what features to omit in the design of a model system. If a concept is a bridge between Nature and science (biology in the present case), a chosen model system is a halfway product between science and technology. Such an approach was often touted as proof of concept.

Since its inception about half a century ago, black lipid membrane or bilayer lipid membrane (BLM) model systems had been explored for such purposes [5–12]. Tien had reported light-induced photovoltage in a BLM made of a chloroplast extract [10,11], which ushered in the study of pigment-containing BLMs as a model system for artificial solar energy conversion. Tien [11] coined the term "photoelectric effects" for the light-induced electric phenomena in biomembranes. He and his colleagues [13,14] subsequently pioneered the construction of a model BLM system, in which the pigment-containing BLM was allowed to separate two aqueous phases that contained electron donors and/or acceptors. In this way, electron transfer reactions were forced to take place across two membrane–water interfaces. Transmembrane charge movements coupled the two interfacial electron transfer reactions together, thus generating a unidirectional photocurrent through the membrane when the BLM was illuminated with light of constant intensity (vectorial electron transport). This model BLM system only partially captured the essence of a photosynthetic membrane, but it vastly simplified the model system and rendered it amenable to rigorous mathematical analysis. We shall call this configuration *coupled interfacial charge transfer (ICT) reactions*. As it turned out subsequently, the charged species to be transported need not be electrons. It could be either protons or other charged species such as chloride ions.

The choice of BLM-based model systems was justified by the fact that naturally occurring photoconverting devices are all membrane based: the photosynthetic apparatus in green plants and in phototropic bacteria comprises hydrophobic proteins embedded in a BLM. The coupled electron transferring BLM, which Tien and his colleagues developed, exhibited a major difference as compared to natural systems. Whereas the small pigments used in the model system are isotropically distributed in the membrane phase, all known natural pigments maintain an asymmetrical distribution in such a way that the pigment or pigment complexes are oriented in a definite direction with respect to the intracellular and extracellular space or with respect to the lumen and the exterior of the photosynthetic organelle. This difference did not seriously compromise the effectiveness of the model system. We shall address this problem in Section 2.

Over the past four decades, a significant number of laboratories explored BLM model systems, and a large variety of pigments had been incorporated into BLM. Most of these studies were cataloged in Tien's two books [15,16] and a major review article [17]. These model systems exhibit a diversity of phenomena but almost all of them could be characterized as coupled ICT BLM systems. Some systems generate transient electric signals, whereas others exhibit prominent sustained photoelectric responses. In spite of the diversity, they share some common features, which are also shared by natural photoconversion membranes. These common features often shed light on the inner working of biological solar energy conversion. General reviews on the photoelectric effects ofbiomembranes are also available [18–23].

Here, we shall examine two pigment-containing BLM model systems. One of them contains small organic pigment molecules, much like the system pioneered by Tien and his colleagues, and the other contains a naturally occurring macromolecular pigment. We shall cross-reference the behaviors of such simple systems with natural systems. Some significant differences are apparent, but, at a deeper level, the operation of both natural and artificial systems converges to the same fundamental concepts and principles. As it turned out, even drastically different natural systems share the same design principle.

2. Engineering Principles

Speaking about solar energy conversion, photodiodes and photosynthetic membranes of green plants easily came to mind. Two aspects are to be considered here: (a) the process of conversion and (b) temporary storages of converted energy in the absence of illumination. Regarding the conversion process, the similarity between a photodiode and photosynthetic membranes is apparent: photoconversion starts with light-induced charge separation. The charge-carrying species in a photodiode are electrons and holes, whereas the charge-carrying species in naturally occurring photoconversion membranes are electron, proton, or even chloride ion. In the case of photodiodes, charge separation is the consequence of light-induced photophysical process, whereas, in the case of biological photoconversion membranes, it is the consequence of light-induced photochemical processes. The intrusion of chemical processes adds to the diversity of phenomenology. In hindsight, biological photoconversion membranes, though relatively complex as compared to a photodiode, are still sufficiently simple to be amenable to rigorous mathematical analysis in terms of elementary physics. It was a reductionist's dream came true. However, the detailed analysis was not only an academic pursuit but also crucial to the understanding of the fundamental process of energy conversion and signal generation.

2.1. Internal Resistance of a Photocell

Depicted in Fig. 1 is a simple schematic diagram, which shows a conventional battery and an electrical device. When the battery is under no load condition (e.g., when the device is disconnected from the battery or when the impedance of the external device is so high that a negligible amount of current flows through the device), an electrical potential difference can be detected across the cathode and the anode of the battery; it is designated as

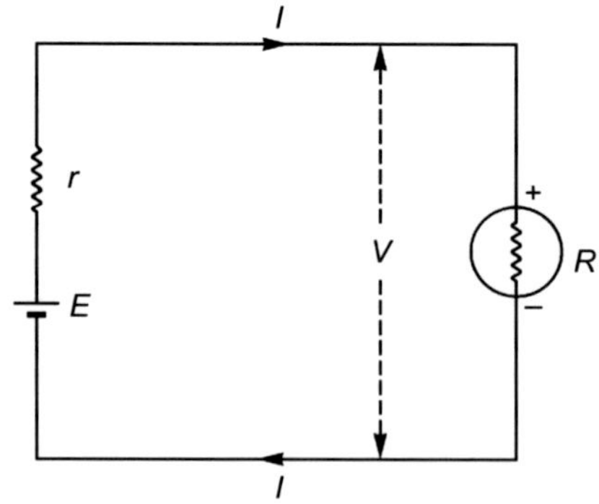

Figure 1 ***A circuit diagram showing a battery and the external load.*** **The circuit diagram represents a battery with emf *E* and internal resistance *r*. The external load (e.g., a lamp, an electric motor, or a complicated electronic device) has resistance *R*. The terminal voltage *V* measured between the cathode and the anode under the loading condition is not exactly equal to *E*, because of internal energy dissipation due to *r*. See text for further explanation.**

the electromotive force (emf), E. When the battery is connected to the external device with a finite resistance R (e.g., an incandescent lamp), a current flows through the circuit according to Ohm's law:

$$I = \frac{E}{R + r}, \tag{1}$$

where r is the internal resistance of the battery. Clearly, the additional nonzero resistance r diminishes the amount of current, I, and reduces the actual voltage, as applied to the device, to a lower value, $V = IR = E - Ir$. In other words, the device never sees the full value of the battery's emf because the flowing current dissipates part of the energy internally by generating heat through the internal resistance. The measured voltage V across the device is the same as the electric potential difference that appears between the two electrodes under the loading condition. It is also known as the *terminal voltage* of the battery. Thus, the energy input into the device becomes

$$VI = I^2R = EI - I^2r = \frac{E^2}{R + r} - \frac{E^2 r}{(R + r)^2} = \frac{E^2}{R\left(1 + \frac{r}{R}\right)^2}. \tag{2}$$

Clearly, minimizing the internal resistance of a battery relative to the external load resistance R increases the efficiency of energy utilization.

2.2. Rectification and One-Way Charge Separation

While energy input is required to separate a pair of charges, their subsequent recombination releases the converted energy. The objectives of a useful design are to avoid internal charge recombination and to let charge recombination take place in the external circuit so that useful mechanical work can be performed. Internal charge recombination, via a reverse (chemical) reaction through the internal resistance r, results in dissipation of the converted energy without doing useful work. This dissipation is in addition to what was mentioned in the previous section during (forward) charge separation. Preventing or minimizing charge recombination in the absence of illumination is a serious design challenge since solar illumination is not expected to be continuous and uninterrupted except perhaps in outer space. Increasing the internal resistance of the photocell slows down the undesirable charge recombination but it also increases the wasteful dissipation as heat during forward charge transfer, as explained earlier.

The above-mentioned dilemma is greatly alleviated in photodiodes because of its inherent nature of rectification: The internal resistance for forward charge separation is low, whereas the resistance for internal charge recombination is significantly higher by nature of the pn junction. Another way to alleviate the dilemma is possible, at least in principle: Maintaining a low internal resistance during illumination but a high internal resistance in the absence of illumination. We shall see later that this latter possibility is not merely wishful thinking.

A one-way charge transfer chemical reaction that goes to completion would automatically fulfill the condition of rectification. However, most chemical reactions never go to completion but, instead, they achieve a chemical equilibrium, that is, an equilibrium between the forward charge transfer and the reverse charge transfer reaction. Usually, the *net* forward charge transfer continues only because of products removal and/or reactant replenishment. Contrary to common belief, rectification is not required for effective energy conversion in a photoconverting membrane. For photoconverting membranes, in which charge separation draws energy from absorbed sunlight, net forward charge separation can be sustained as long as illumination continues. Because of the continuous input of light, chemical equilibrium is never achieved. Instead, a steady state is established if the influx of light is of constant magnitude. In other words, the process of charge separation is "pumped" by absorbed light energy. Rectification is therefore not a prerequisite for effective photoconversion to take place. We shall demonstrate this feasibility with a model BLM system (Section 3).

2.3. Spatial Anisotropy of Charge Separation

For photoconverted energy to be useful, light-induced charge separation must be spatially anisotropic. It would not work in a homogeneous solution phase, since charge separation would proceed randomly in all directions (spatially isotropic). An interface between unlike materials provides a natural environment for anisotropic charge separation to take place, the pn junction of a photodiode being one of the most familiar examples. The situation is similar in biological solar energy conversion. All known photosynthetic apparatuses are membrane based. All of them convert absorbed light energy into charge separation across the entire span of a thin membrane of about 100 Å in thickness. Obviously, the highly insulating lipid membrane prevents separated charges from recombining internally. Therefore, light-induced charge movement in biomembranes must cross two interfaces, instead of just one. All these membranes exhibit a functional asymmetry: charges are moving from one side of the membrane to the other, known as *vectorial charge transport*. The functional asymmetry is made possible by virtue of asymmetric orientation of the membrane-bound pigment complexes with respect to the lipid bilayer membrane. In model BLM systems, the functional asymmetry must be maintained. As evident in the original system analyzed by Tien and his colleagues, as well as many subsequent studies, the asymmetry arose from asymmetrical distributions of aqueous-borne electron donors and acceptors. The two examples to be analyzed in this chapter reflect this fundamental difference.

2.4. Methods of Measurement

The BLM model system, though highly simplified, is still a complex system. In the early stage of BLM studies in which light was not involved in the process of charge movements, the methodology was directly transplanted from electrophysiology with considerable success. In the studies of photoelectric effects in model BLM systems, electrophysiological techniques (and the corresponding electrochemical techniques), such as the voltage-clamp (potentiostat) and the current-clamp (galvanostat) methods, were also utilized. However, new challenges appeared. One needed to jump out of the proverbial box of classical electrophysiology.

The development and utilization of the voltage-clamp method were crucial to the landmark achievement of Hodgkin and Huxley in the mid-twentieth century [24,25]. By applying the voltage-clamp method of measurements, they analyzed isolated nerve membranes in terms of equivalent (electric) circuit (called equivalent circuit analysis). Note that an equivalent circuit comprises *discrete* electric elements, such as emfs (or batteries), resistances, and capacitances. But the electric counterparts of these elements in a real membrane are *continuous* and *distributed*. Therefore, an equivalent circuit never fully represents a real membrane, but it captures sufficient features to be useful in analysis. In other words, the representation is not an isomorphism but rather a homomorphism. Hodgkin and Huxley were able to use this approach to elucidate the nerve excitation mechanism without being misled by the pitfall.

Basically, Hodgkin and Huxley applied Ohm's law to nerve cell membranes. They considered three parameters: emf, membrane current, and membrane resistance (or its reciprocal, conductance). If one specifies and fixes the membrane voltage/emf as an independent variable and measures the membrane currents, Ohm's law yields the computed value of membrane resistance or, as electrophysiologists preferred, membrane conductance (voltage-clamp measurement). Alternatively, if one specifies and fixes the membrane current as an independent variable and measures the voltage, Ohm's law yields the computed value of membrane conductance (called current-clamp or galvostat method). In the special cases in which the membrane current is set at zero, the measurement is also known as an open-circuit measurement. It corresponds to the no load condition, mentioned in the previous section.

The complexity in classical electrophysiology stems from the experimental fact that the membrane conductance (or resistance) is not constant (i.e., non-ohmic); its value varies during the process of nerve excitation. Worse yet, the membrane conductance comprised several subsystems: Na^+ current, K^+ current, etc.; each of them exhibits a different way of being non-ohmic. The most crucial insight of Hodgkin and Huxley was that the membrane conductance and the corresponding conductances of the subionic systems are voltage dependent as well as time dependent [24]. The identification of transmembrane voltage as the independent variable of the system and its subsystems contributed to the success of their voltage-clamp measurements. The deployment of equivalent circuit analysis elevated biology from a science of qualitative descriptions and explanations to quantitative science, which used to be monopolized by physics and chemistry and their nonliving progenies. Needless to say, the possibility to make quantitative predictions and subsequent experimental verifications greatly enhanced the likelihood of convincing an initially skeptical audience and that of eliminating a seemingly correct but ultimately false model.

Earlier studies of photoelectric BLM systems followed classical electrophysiology as a precedent for guidance. However, it did not take long for us to realize that we had encountered a new complication: the action of light, which generates additional electrical parameters. Before we could analyze the system in a meaningful manner, we needed to determine what additional parameters must be included in the equivalent circuit analysis. In pigment-containing BLM systems, illumination usually causes an increase in the transmembrane current in the case of voltage-clamp measurements, or an increase in the transmembrane voltage in the case of current-clamp measurements. These additional currents or voltages caused by the action of light were often loosely referred to as photocurrents or photovoltages. The intrusion of additional parameters also made it clear that a single measurement of either voltage clamp or current clamp is insufficient to determine all the unknown parameters; a minimum of two independent measurements are needed to determine all electrical parameters in the equivalent circuit description. In the absence of the "other" measurement, there are two possible sources that contribute to the increase of measured currents: light may generate additional emf (photoemf) and/or additional conductance (photoconductance). These complications were not anticipated in classical electrophysiology. But the recognition of the missing "unknown" measurement method converted a blind trial and error to a more directed search of new method of measurement (heuristic search).

The presence of two interfaces in a biomembrane, as compared to photodiodes, is yet another complication. A photodiode has one active interface where charge separation occurs: the pn junction. Light-induced charge separation in a BLM system (or any membrane-based model systems), in which coupled ICTs occur, involves two interfaces: the two membrane–water interfaces, which productive charge separation must traverse. Moreover, the two interfaces are not too far apart; a small distance of about 100 Å separates the two membrane–water interfaces. This small distance is on a spatial scale too small to be considered macroscopic and too large to be considered microscopic. The spatial scale was often referred to as *mesoscopic*. As we shall see, the latter feature has a significant impact on interpretation of electrical data, which was not foreseen in classical electrophysiology, and which was a source of confusion in the literature of photoelectric effects in biomembranes and model membranes. The ubiquitous presence of reverse chemical reactions added an additional complication, since ICT reactions are not exempted from the complication of reverse reactions. At least part of the confusion that occurred in the literature was caused by the *undeclared and unproven assumption* that interfacial charge movements are rectified, just like photodiodes, and reverse reactions were thus unjustifiably excluded from consideration. However, the presence of reverse charge transfer reactions makes it possible to observe photoelectric signals like alternating electric currents. This seemingly remote possibility becomes a reality because of the ultrathinness of a BLM. The close proximity of the two interfaces allows the electrical event at one interface to influence the electrical event at the other interface, and vice versa. In other words, photoelectric signals are *AC coupled* across the membrane. We therefore need to distinguish between two types of photoelectric signals.

Constant illumination eventually leads to a steady state, in which a constant photoelectric current flows through the membrane. It is similar to direct currents, which we encountered in conventional electricity. We shall refer to it as the *DC photoelectric effect*, and the measured steady-state photocurrent shall be called a *DC photocurrent*. Often, the system behavior could be better understood if the experimenter probes the kinetics of the model system with a brief light pulse to elicit a transient photocurrent. As we shall demonstrate, the observed photosignals are predominantly transient signals much like conventional AC currents. We shall refer to the measured photocurrent as an *AC photocurrent*. We shall refer to the corresponding phenomenon as the *AC photoelectric effect*. In hindsight, it should not have had to take a pulsed light experimental regime to bring our attention to the possible presence of AC photoelectric signal. A steady-state measurement of DC photoelectric signals often included transient spikes at the onset of (square wave) illumination and at the end of illumination. However, humans' habitual assumption that biological systems are inherently too complex to be amenable to simple explanations steered investigators away from seeking simple explanations instead. Perhaps biologists need to be constantly reminded of Ockham's razor.

3. A Simple Pigment-Containing Model BLM System

The first step of our analysis is to establish a theoretical framework that allows for a mechanistic understanding of photosignal generation. The analysis presented in the previous section serves to remind us that what we were to analyze was neither a pure electrical system nor a pure photochemical system since interactions between electrical and chemical events were to be expected. To address the essence of this new feature, we must choose a bare-bone model system that still captures the essence of coupled ICT reactions in a BLM. We thus initially shied away from using naturally occurring pigments. We must implement an uncomplicated charge transfer reaction in the BLM environment. From previous works of Fuhrhop and Mauzerall [26,27], it was known that magnesium porphyrins, such as magnesium mesoporphyrin IX and magnesium octaethylporphyrin (MgOEP), undergo *reversible one-electron* redox reaction, when these pigments are coupled to a pair of redox reagents, such as potassium ferricyanide and potassium ferrocyanide.

$$\mathrm{P} + \mathrm{Fe(CN)}_6^{3-} \rightleftarrows \mathrm{P}^+ + \mathrm{Fe(CN)}_6^{4-},$$

where P and P^+ are the ground state neutral magnesium porphyrin and the corresponding oxidized monocation, respectively. It turned out that the oxidized porphyrin monocation P^+ is *thermodynamically* stable. The photochemistry is thus simple and devoid of unwanted side reactions. Thus, we had a simple light-pumped photochemical reaction cycle:

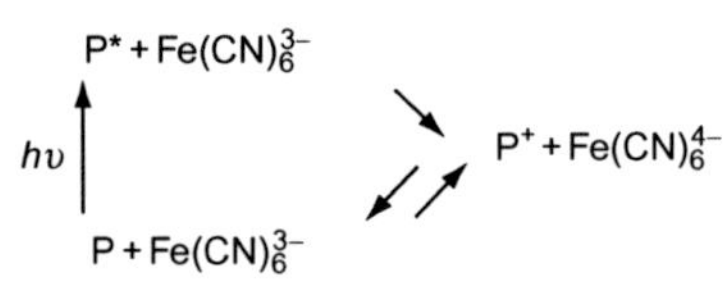

where P★ is the neutral magnesium porphyrin in the excited state. As we shall see, the second system to be analyzed in this chapter is also featured with a more complex photochemical cycle or, in brief, photocycle: light pumps a cyclic regime of charge transfer reactions. When the photocycle is implemented in an environment of membrane–water interfaces, a solar cell is formed.

To configure the magnesium porphyrin photocycle to serve the purpose of solar energy conversion, magnesium porphyrins must be incorporated into the BLM so that light-driven electron transfer reactions take place across the membrane. A simple partial organic synthesis renders magnesium porphyrins hydrophobic so as to be confined to the membrane phase. Light-driven electron transfer reactions could then take place between the membrane-bound pigment and the aqueous-borne redox reagents, potassium ferricyanide, and potassium ferrocyanide. By adding a long-chain alcohol, *n*-amyl alcohol, to its two carboxylic side chains, magnesium mesoporphyrin IX can be esterified and rendered sufficiently hydrophobic. MgOEP is naturally hydrophobic and it was used without further modification.

As shown in Fig. 2, P, P★, and P^+ are membrane bound, whereas $\mathrm{Fe(CN)}_6^{3-}$ and $\mathrm{Fe(CN)}_6^{4-}$ are hydrophilic and are therefore confined to the aqueous phases. The segregation of the participating reactants forces

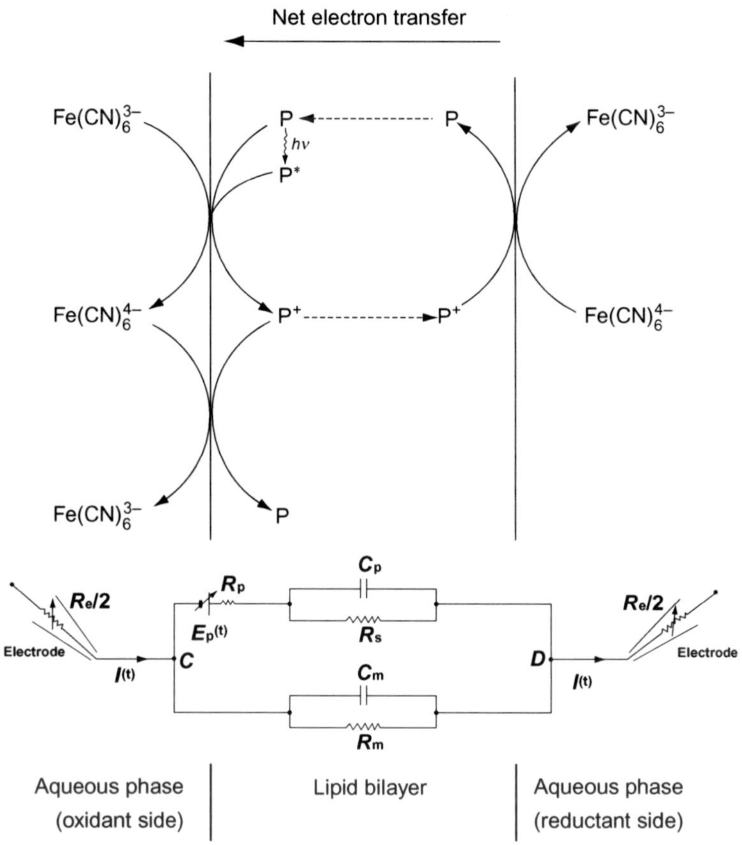

Figure 2 *A simple light-driven electron pump.* The bilayer lipid membrane (BLM) contains a lipid-soluble magnesium porphyrin and separates two aqueous phases with asymmetric compositions of potassium ferricyanide (electron acceptor) and potassium ferrocyanide (electron donor). The symbol P stands for the ground state pigment, P* for the excited state pigment, and P^+ for the oxidized form of the pigment (monocation). P and P* are electron donors and P^+ is the electron acceptor. Interfacial electron transfer reactions occur between the membrane-bound pigment and the aqueous-bourne electron donor/acceptor. Electrons are preferentially transported from the reductant side (where the donor concentration is higher than the acceptor concentration) to the oxidant side (where the acceptor concentration is higher than the donor concentration). Diffusion of P and P^+ across the membrane enables the coupling the two separate interfacial electron transfer reactions, thus generating a net transport of electrons from the donor side to the acceptor side. The reverse reaction at the oxidant interface is also shown. The equivalent circuit is shown at the bottom. The photoemf $E_p(t)$ with its internal resistance, R_p, is connected in series with chemical capacitance, C_p, and transmembrane DC resistance, R_s. C_p provides the pathway for the AC photocurrent. R_s is the resistance encountered by the DC photocurrent. The latter is a function of mobility of P and P^+ inside the membrane. The ordinary membrane (ionic) resistance, R_m, and capacitance, C_m, are placed in parallel with the photoemf. The access impedance (resistance), R_e, includes the amplifier input impedance, the electrode impedance, and the impedance of the intervening electrolyte solution. Adapted from Ref. [32].

the previously described electron transfer reactions to take place across membrane–water interfaces. In fact, there are two interfacial electron transfer reactions, one at each membrane–water interface. These two reactions are *chemically coupled* by means of diffusion of P★ and P^+ in the transmembrane directions. The electron transfer reactions could be made to take place preferentially from one side of the membrane toward the other side, if the redox reagents are distributed asymmetrically in the two aqueous phases. A DC current could be detected and it represented the net vectorial charge transport across the membrane. Routinely, one aqueous phase (called the oxidant side or the electron acceptor side) contains predominantly ferricyanide, $Fe(CN)_6{}^{3-}$, and the other aqueous phase (called the reductant side or the electron donor side) contains predominantly ferrocyanide, $Fe(CN)_6^{4-}$. The energy that drives the electron transport in the absence of illumination was previously stored as a transmembrane gradient of $Fe(CN)_6{}^{3-}$ and $Fe(CN)_6^{4-}$. Because of the aqueous asymmetry, there exists a concentration gradient of P^+: its concentration is higher on the oxidant side than on the reductant side. A similar concentration gradient of system of P of opposite polarity also exists. In the absence of illumination, the transmembrane chemical potential of the ferricyanide/ferrocyanide redox couple drives the transmembrane migration of neutral pigment P and monocationic porphyrin P^+ in opposite directions, thus resulting in the generation of a transmembrane electric potential difference. An equilibrium is established when the transmembrane electrical potential balances the transmembrane chemical potential, similar to what transpires in the formation of a diffusion potential in nonpigmented BLM. In the absence of illumination, a dark emf and a dark steady current can be measured under open-circuit or short-circuit conditions, respectively. The dark electric phenomenon reflects the redox reactions implemented in a three-phase environment in accordance with the reversible one-electron transfer reaction: $P + Fe(CN)_6^{3-} \rightleftarrows P^+ + Fe(CN)_6^{4-}$. Needless to say, the uncomplicated photochemical regime vastly simplified the interpretation of photoelectric signals. However, the dark electric phenomenon is not what we set out to study. Rather, they represent the background signal, on top of which photoelectric events take place.

Under illumination with a constant light source, forward electron transfer from P★ to $Fe(CN)_6^{3-}$ is much enhanced, and the above-mentioned photocycle is thus kicked into action. When this enhancement takes place in a membrane system configured as in Fig. 2, an additional steady current can be measured under short-circuit conditions. This additional component is defined as the *DC photocurrent.* The experimentally observed DC dark current and DC photocurrent have the same polarity that is consistent with a net electron movement from the reductant side to the oxidant side. That is, the DC current flows from the oxidant side across the membrane to the reductant side. So far, we have just considered steady-state electrical behavior of the model system. When a constant light source is suddenly turned on, a transient signal precedes the steady-state photocurrent. Likewise, another electric transient follows the steady-state photocurrent when the constant illumination is suddenly terminated. These transient currents are intimately related to the AC photoelectric effect. But the characteristic "on" spikes and "off" spikes look like the response of an electric differentiating circuit, thus inspiring the designation of *differential responsivity* to the collective behavior of photoelectric phenomenon in the literature.

3.1. The Significance of Reverse Electron Transfers and the AC Photocurrent

At first glance, the transient currents mentioned above looks remarkably similar to the transient signals when a square pulse of electric signal passes through an *RC* circuit such as a high-pass filter (a differentiating circuit). In the literature of photoelectric biomembranes, these transient electric signals were thought to reflect the interaction with the ever-present membrane capacitance arising from the insulating dielectric of phospholipid bilayers. In reality, it turned out to be more complicated than that. For nearly a quarter century, the literature was filled with conflicting interpretations and paradoxes. The version to be presented here is perhaps the simplest and yet still self-consistent interpretation. But that is for the readers to judge by themselves. Readers are also referred to some key references that advocated alternative views [20,28,29].

It can be shown rigorously that the transient electric phenomenon reflects the reverse electron transfer reactions at the two membrane–water interfaces, especially the oxidant side [18]. That this transient signal is readily detectable by an externally located current-measuring device is a consequence of ultrathinness of the membrane; the signal is AC coupled through the membrane. In addition to the forward electron transfers taking place at the two membrane surfaces, there exist two reverse electron transfer reactions, one each at the two membrane–water interfaces, which tend to drive the electron movement in the opposite direction. From the point of view of photon energy conversion, only net forward electron transfers are considered productive. These reverse electron transfer reactions are counterproductive, since forward charge transfers are, at least, partially canceled by subsequent reverse transfers. The DC photocurrent reflects the difference between forward charge transfers and reverse charge transfers. The reverse reactions take place even during illumination (implying the lack of rectification). However, as long as illumination is maintained, the system will reach a steady-state and a steady-state photocurrent can be observed. The steady-state photocurrent is sustained by a continuous input of light energy, that is, the electron transfer system is being pumped by light energy. As we shall demonstrate experimentally, the magnesium porphyrin–BLM system does not exhibit rectification. Therefore, rectification, while desirable, is not a prerequisite of photon energy conversion.

The reverse reactions make their prominent presence when the membrane is illuminated with a brief (microsecond) pulsed light. Upon illumination with such a pulsed light, a transient photocurrent can be detected [30]. This transient photocurrent flows initially in the same direction as the DC photocurrent but subsequently reverses its polarity. For reasons explained elsewhere [18,19], the microsecond transient photocurrent reflects predominantly the reverse electron transfer reactions; the steady current that reflects the net electron transfer is dwarfed in amplitude by comparison with the transient current. This transient photocurrent turned out to be the *AC photocurrent* alluded to in the previous section. An AC photocurrent results in no net charge transfer, since the amount of charge recombination cancels the amount of charge separation completely. If, however, a steady illumination lasting longer than milliseconds is applied, the transient photocurrent becomes much smaller. The reverse electron transfers may still show up in the photocurrent recording if a steady light of constant intensity is suddenly turned on or turned off. The superficial difference in waveform of the pulsed light-induced photosignal and the steady light-induced photosignal can be reconciled in terms of the equivalent circuit shown in Fig. 2. The detailed explanation will be presented next.

3.2. Equivalent Circuit Analysis

It has been established rigorously [18] that the measured photocurrent can be simulated by an equivalent circuit shown in Fig. 2, in which the driving potential, the photoemf, $E_p(t)$, follows exactly the time course of the illuminating light pulse. Essentially, $E_p(t)$ drives a current through the membrane that is AC coupled via the RC circuit comprising R_p and C_p. The combination of R_p and C_p essentially forms a high-pass filter circuit. That is why photocurrent elicited by a microsecond light pulse contains predominantly AC components. The presence of R_s allows a DC photocurrent to go through the membrane, but this component is not AC coupled and is not preferentially amplified at high frequencies (i.e., microsecond time scale). Thus, not all transiently separated charges subsequently recombined. This portion of photocurrent that manages to leak through the membrane represents a net charge transfer and a net light energy conversion.

The introduction of the capacitative element C_p inadvertently elicited a controversy that lasted for almost a quarter century. A significant number of investigators claimed that C_p was actually the ordinary membrane capacitance C_m in disguise and that C_p was an algebraic artifact. We provided proof that it is impossible to reduce the capacitative element C_p to the ordinary membrane capacitance C_m, on the basis of charge distribution patterns on the fundamental capacitance elements, known as geometric capacitance and diffuse double layer capacitance [31]. Thus, C_p and C_m are *physically distinct* entities. This new capacitance stems from interfacial electron transfers. We named it *chemical capacitance.* An equivalent circuit that incorporates this capacitance is shown in the bottom part of Fig. 2. The interfacial resistance, R_p, can also be regarded as the internal resistance of the photocell. Also included is the transmembrane resistance, R_s, which provides the transmembrane pathway for the pair of electron shuttles, P and P^+. Circuit analysis indicates that the photocurrent decays in a single exponential term:

$$I(t) = \frac{E_p(t)}{R_p} - \left(\frac{1}{\tau_p} - \frac{1}{R_p C_p}\right) \int_0^t \frac{E_p(u)}{R_p} \exp\left(\frac{u-t}{\tau_p}\right) du, \tag{3}$$

where

$$\frac{1}{\tau_p} = \frac{1}{R_p C_p} + \frac{1}{R_s C_p}. \tag{4}$$

If the light pulse is sufficiently brief, it can be represented mathematically by a δ- function, thus yielding a transient photoemf: $E_p(t) = E_0\delta(t)$. The photocurrent becomes

$$I(t) = \frac{E_0}{R_p}\left[\delta(t) - \left(\frac{1}{\tau_p} - \frac{1}{R_p C_p}\right) \exp\left(-\frac{t}{\tau_p}\right)\right]. \tag{5}$$

The signal rises as a sharp peak and then reverses its polarity immediately; it subsequently decays with a single exponential time constant.

3.3. Effect of Access Impedance

It turned out that a voltage-clamp measurement of the photosignals was not so straightforward. In principle, a short-circuit condition can be met if the input impedance of the current-measuring device is smaller than the source impedance of the membrane. For a BLM membrane without an incorporated pigment, an electrode resistance of a few kiloOhms is considered negligible, since a BLM membrane possesses a high source impedance of the magnitude of the order of megaOhms.

A light-sensitive BLM deserves special attention. The presence of C_p drastically diminishes the "internal" impedance of the membrane (source impedance). As a consequence, an electrode resistance of 20 kΩ (that of a pair of calomel electrodes) renders the measurement condition *virtually open-circuit*, since the source impedance drops dramatically to about the same order of magnitudes at the megaHertz frequency range. Thus, the electrode resistance is hardly negligible, and the membrane was not effectively short-circuited right at the membrane surfaces. The impedance that is located between the two membrane surfaces and the two device-input terminals is called the *access impedance*, which is depicted as R_e in Fig. 2. The electrode resistance is the major contributor to the access impedance. The intervening solution between the electrode tips and the membrane surfaces also contributes to part of the access impedance. At the submicrosecond range, even an access impedance of 380 Ω is not negligible.

The tricky part was that the investigator would not be aware of this pitfall until they realized that C_p exists. It was difficult to suspect a heretofore unknown entity as the culprit of a variety of inexplicable inconsistencies that subsequently appeared in the literature. Of course, once suspected, the inclusion of C_p in the equivalent circuit resolved almost all puzzling "paradoxes." While it is difficult to eliminate the influence of the access impedance, its effects can be readily included in the analysis.

The price to pay in a measurement with nonzero access impedance is that the near-short-circuit current so measured is distorted by interactions of the *RC* circuit. The experimental condition is not strictly a short-circuit one (voltage clamped). It is better to refer to the measurement as a *near-short-circuit* measurement. We found experimentally that the relaxation of the AC photocurrent varies with the electrode resistance. In other words, the access impedance is modulating the photosignals, and the true short-circuit signals are not directly measured.

By deliberately adding additional series resistance to the existing access impedance, we found that the observed AC photoelectric signal to be profoundly affected by the change of access impedance. Thus, the two exponential decays cannot and should not be directly interpreted as the underlying chemical relaxations. The common practice of directly interpreting the measured relaxation kinetics as the true ones contributed to confusion that stretched over a quarter century. The true time course of the photocurrent could only be recovered by deconvolution of the measured signals (the process of distorting the signal by means of a prescribed algorithm is mathematically called a *convolution*). The resulting photocurrent now decays in two exponential terms,

$$I(t) = \frac{\frac{1}{\tau_s} - \frac{1}{R_s C_p}}{R_e C_m \left(\frac{1}{\tau_s} - \frac{1}{\tau_l}\right)} \int_0^t \frac{E_p(u)}{R_p} \exp\left(\frac{u-t}{\tau_s}\right) du - \frac{\frac{1}{\tau_l} - \frac{1}{R_s C_p}}{R_e C_m \left(\frac{1}{\tau_s} - \frac{1}{\tau_l}\right)} \int_0^t \frac{E_p(u)}{R_p} \exp\left(\frac{u-t}{\tau_l}\right) du, \tag{6}$$

where τ_s and τ_l are defined by the following equations:

$$\frac{1}{\tau_s}=\frac{1}{2}\left[\left(\frac{1}{R_pC_m}+\frac{1}{\tau_p}+\frac{1}{\tau_m}\right)+\sqrt{\left(\frac{1}{R_pC_m}+\frac{1}{\tau_p}+\frac{1}{\tau_m}\right)^2-4\left(\frac{1}{R_pC_mR_sC_p}+\frac{1}{\tau_p\tau_m}\right)}\right], \tag{7}$$

$$\frac{1}{\tau_l}=\frac{1}{2}\left[\left(\frac{1}{R_pC_m}+\frac{1}{\tau_p}+\frac{1}{\tau_m}\right)-\sqrt{\left(\frac{1}{R_pC_m}+\frac{1}{\tau_p}+\frac{1}{\tau_m}\right)^2-4\left(\frac{1}{R_pC_mR_sC_p}+\frac{1}{\tau_p\tau_m}\right)}\right]. \tag{8}$$

The parameter τ_p is defined in Eq. (4), and the parameter τ_m is defined by the following equation:

$$\frac{1}{\tau_m}=\frac{1}{R_eC_m}+\frac{1}{R_mC_m}. \tag{9}$$

By setting R_s to infinity, the DC component becomes zero. Again, given a δ-function-like impulse of illumination, the observed AC photocurrent will decay in two time constants, τ_s and τ_l.

$$I_{AC}(t)=\frac{E_0}{R_pR_eC_m\left(\frac{1}{\tau_s}-\frac{1}{\tau_l}\right)}\left[\frac{1}{\tau_s}\exp\left(-\frac{t}{\tau_p}\right)-\frac{1}{\tau_l}\exp\left(-\frac{t}{\tau_p}\right)\right]. \tag{10}$$

Note that Eq. (10) consists of two exponential terms of opposite polarity. The two time constants, τ_s and τ_l, are neither τ_p nor τ_m but rather a mixture of a number of RC constants including τ_p and τ_m. However, the intrinsic relaxation time constant τ_p can be recovered by deconvolution. In fact, the pair of simultaneous Eqs. (7) and (8) can be solved for R_p and C_p in terms of the remaining parameters. These remaining parameters are all experimentally measurable. It can be shown that unique solutions for R_p and C_p do exist [18]. The scale factor E_0 can be determined by normalization of the computed waveform against the measured one. Thus, the equivalent circuit is completely defined by experimental measurements, and there are no freely adjustable parameters. Figure 3 shows both the actually measured signal and its agreement with the computed signal.

The full equivalent circuit predicts that the time course of the observed photosignal is strongly dependent on the access impedance, R_e (Fig. 4). Under short-circuit conditions when R_e is much smaller than the source impedance, all the photocurrent proceeds to the external measuring device.

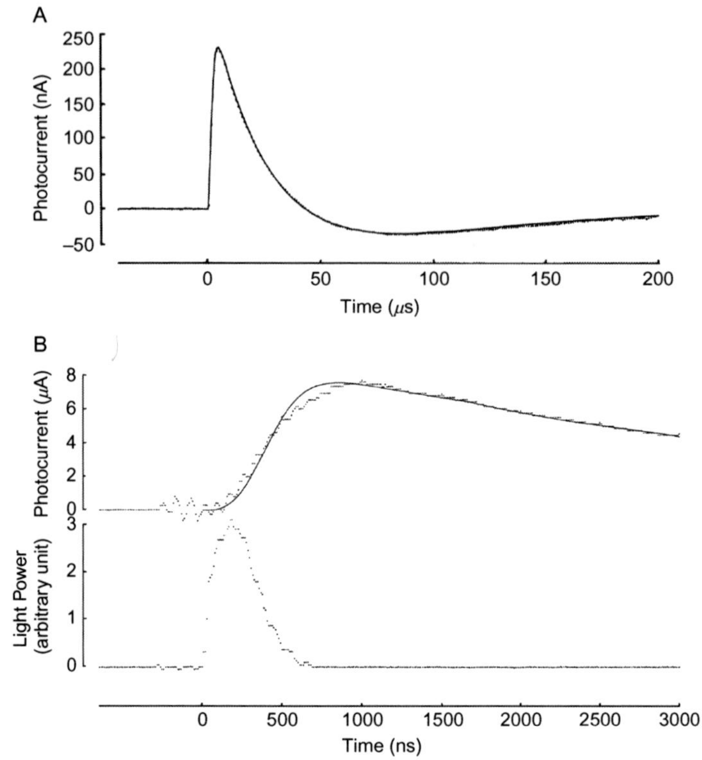

Figure 3 *Comparison of experimentally measured and computed photoelectric responses.* The BLM contains about 5 mM magnesium mesoporphyrin IX di-*n*-amyl ester. Both aqueous phases contain 1 M NaCl and 10 mM phosphate buffer at pH 7.2. In addition, the oxidant side contains 20 mM potassium ferricyanide and 0.5 mM potassium ferrocyanide, and the reductant side contains 20 mM potassium ferrocyanide (the omission of potassium ferrocyanide at the reductant side has no major effect on the transient photoelectric signal). The access impedance is 5.1 kΩ in Record A and 380 Ω in Record B. The instrumental time constant is 1.5 μs in Record A and 150 ns in Record B. The photoemf follows the time course of the power (intensity) of the dye laser pulse (output 590 nm). The amplitude of the photoemf is obtained by normalization of the peak. All the input parameters used for computation are experimentally measured. There are no freely adjustable parameters except the normalization factor. Adapted from Refs. [32] and [33].

The photosignal decays in a single exponential time constant, τ_p, which is a function of C_p, R_p, and R_s but not of R_m and C_m. To a good approximation, τ_p is equal to R_pC_p, since R_s is much greater than R_p. Under open-circuit conditions when R_e is much greater than the source impedance, all the photocurrent proceeds to charge C_m and the photosignal relaxation will then be dominated by the membrane RC relaxation with a decay time

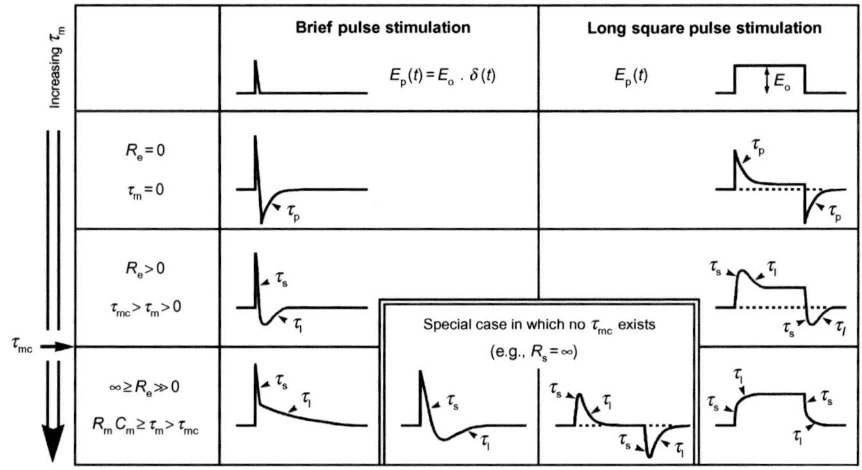

Figure 4 *The photosignal relaxation time courses are calculated in accordance with the equivalent circuit shown in Fig.* 2. **Two types of illumination are considered: brief δ-function-like light pulse and long square-wave light pulse. Note that the relaxation time course changes with the change of the discharging time constant, τ_m, as defined by Eq. (9). Under open-circuit conditions (i.e., $R_e = \infty$), the photosignal relaxes in two exponential time constants, τ_s and τ_l, of which the longer one, τ_l, approaches the membrane *RC* relaxation time constant, $R_m C_m$. Under short-circuit or near-short-circuit conditions, the photosignal generated by a long square light pulse exhibits an "on" spike and an "off" spike, as expected for a linear high-pass *RC* filter. This characteristic waveform may disappear if τ_m exceeds a critical value, τ_{mc}. Adapted from Ref. [19].**

constant of $R_m C_m$. However, when the access impedance is matched to the source impedance, the photocurrent splits into two branches: one branch proceeds to charge C_m and the other flows to the external detector (either a voltage or a current meter). The photosignal relaxation then reflects the interaction of the two *RC* networks and decays in two exponential time constants, which are neither $R_p C_p$ nor $R_m C_m$. The time constant τ_p can be recovered by deconvolution, using only parameters that can be determined experimentally. Contrary to the common practice of curve fitting with a number of adjustable parameters, the present scheme has no freely adjustable parameters. We observed that when R_e is comparable to the source impedance of the membrane, data obtained by deconvolution are more accurate than when R_e is at its extreme values (i.e., much greater or much smaller than the source impedance). In fact, individual values of R_p and C_p can be so determined, whereas in a short-circuit measurement, only the product, $R_p C_p$ (but not the individual values of R_p and C_p), can be determined. We routinely "tuned" the value R_e in order to optimize the measurements of R_p and C_p. For this reason, this method is named the *tunable voltage-clamp method* [32,33].

The dramatic transformation of photosignals when the duration of illumination was progressively increased from a brief light pulse to a steady square-wave pulse was demonstrated by photoelectric data taken from a cyanine dye-containing BLM, with an electron donor present in one of the two aqueous phases [34] (Fig. 5). The complicated transient waveform at the onset and the cessation of illumination can be understood in terms of the same equivalent circuit presented in Fig. 2.

Here, a technical comment is in order. Superficially, a voltmeter is for measuring open-circuit voltages, and an ammeter is for measuring short-circuit currents. In measuring electric signals in a BLM system, especially a pigment-containing BLM system, the task is far trickier than expected by an unsuspicious investigator (ourselves included before we recognized the pitfall), and an additional precaution must be kept in mind. Thus, a regular voltmeter cannot be used to measure membrane voltage because the BLM system has inherently high (source) impedance. An electrometer (with input impedance as high as 10^{16} Ω) is required. Likewise, a picoammeter might not be suitable for measuring short-circuit currents in a pigment-containing BLM, since the source impedance might be unexpectedly low. An intended short-circuit measurement using a commercial picoammeter (input impedance about 100 kΩ) could bring about a measurement condition that was actually closer to open-circuit than to short-circuit conditions because the source impedance at high frequency could be easily reduced to the order of

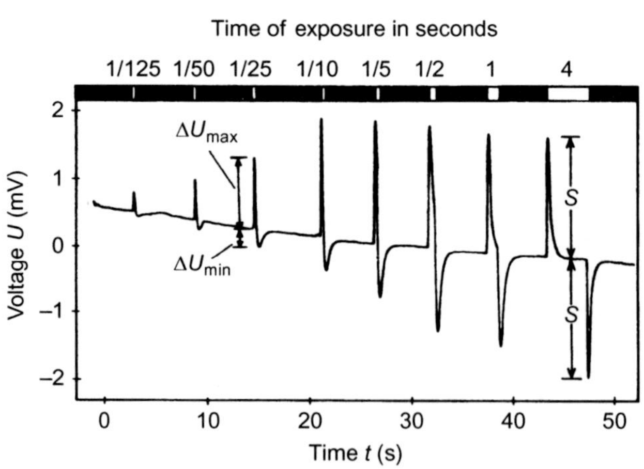

Figure 5 *Effect of changing the duration of the stimulating square-wave light pulse on the time course of a photoelectric signal.* A cyanine dye was adsorbed to one side of a planar BLM. An electron doner was present in the same side of the aqueous phase. The successive photovoltage responses to light pulses of increasing duration (1/125, 1/50, 1/25, 1/10, 1/5, 1/2, 1, and 4 s) are shown. The progressive waveform changes are consistent with the equivalent circuit of Fig. 2, with $R_e = \infty$. Adapted from Ref. [34].

10 kΩ (that of R_p). This was exactly the case of measurements found in Ref. [35]. There was a disturbing practice of citing the agreement with the following equation as proof of validity of measurements.

$$I_{sc}(t) \propto \frac{\mathrm{d}V_{oc}(t)}{\mathrm{d}t}, \tag{11}$$

where I_{sc} is the measured short-circuit current and V_{oc} is the measured open-circuit voltage. As explained elsewhere (Section 11.2 of Ref. [22]), the time course of a measured short-circuit current never becomes proportional to the first derivative of its corresponding open-circuit current, unless one or both of them have been measured under incorrect conditions. The agreement was telltale evidence for a technical failure: the short-circuit measurement shown in Ref. [35] was actually carried out under near-open-circuit condition. The culprit was the improper use of a picoammeter in measuring short-circuit currents in a light-sensitive BLM.

The concept of chemical capacitance also offered a simple explanation why a pulsed light-induced photosignal measured in high-frequency ranges consists of a negligible contribution from the DC photocurrent. The DC photocurrent was not absent but it was small compared to the magnitude of noise when the AC photocurrent is properly measured. The built-in high-pass filter consisting of R_p and C_p selectively suppresses slower electrical signals and preferentially let faster ones pass through without attenuation. This is precisely what a high-pass filter is supposed to do: suppress low-frequency signals.

4. Bacteriorhodopsin: A Natural Light-Driven Proton Pump

Next, we shall examine a photoconverting BLM system that possesses the next level of complexity. In this case, the membrane-bound pigment being used is called bacteriorhodopsin (bR), which was discovered by Oesterhelt and Stoeckenius in 1971 [36,37]. bR is the only protein component of the purple membrane patch of *Halobacterium salinarum* (formerly *Halobacterium halobium*). Its name reflects its chemical similarity to the visual pigment rhodopsin; both contain a covalently bound vitamin A aldehyde as the chromophore, and both consist of seven segments of transmembrane α-helices with *C*-terminus located on the cytoplasmic side and the N-terminus on the opposite side. Its photochemistry is characterized by a cyclic photochemical reaction scheme (photocycle). The cyclic reaction scheme is pumped by light, resulting in the transport of protons from the cytoplasmic side of the purple membrane to the extracellular side. bR is one of the best understood ionic pumps (for general information, see Refs. [38–48]). In contrast to the rather labile pigment rhodopsin, bR is exceptionally stable. Over the past few decades, bR was one of the most popular biomaterials to be exploited for biomolecular sensor research and development. Interested readers are referred to Refs. [49–57].

Rhodopsin and bR exhibit similar AC photoelectric signals in their native membranes. The early receptor potential (ERP)—the AC photoelectric signal in a visual membrane—consists of two electric components: a faster *R*1 component that is temperature insensitive and a slower *R*2 component that can be reversibly inhibited by low temperature (0 °C) [58,59]. The similarly behaved ERP-like signal components in the purple membrane were named *B*1 and *B*2, respectively [60].

There was a long-standing controversy regarding the molecular mechanism of the ERP. Our concurrent studies of the Mg-porphyrin-containing BLM and reconstituted bR membranes helped elucidate the molecular mechanism in a unique way because these studies led to insights that would be hard to come by had we studied only one of them but not the other. Historically, the ERP has been interpreted as the electrical manifestation of light-induced intramolecular charge displacement in rhodopsin. An examination of the AC photocurrent in the Mg-porphyrin BLM system reveals that the signal shares all the major characteristics of the ERP (Table 1 in Ref. [18]). Most notably, ERP was found to be an AC signal that satisfies the zero time-integral condition [61], which is a hallmark of almost purely AC photosignals. This suggests that an ICT can be an alternative mechanism for ERP generation; an intramolecular charge displacement is not the only possibility. In addition, the two mechanisms are not mutually exclusive, in principle.

It has long been known that the ERP *R*2 component is time-correlated with the spectroscopic signal of the metarhodopsin I to metarhodopsin II reaction [59], which involves the binding of a proton from the (cytoplasmic) aqueous phase [62,63] and which is one of the key steps of rhodopsin photobleaching reactions, leading to visual transduction [64]. Thus, in principle, the *R*2 signal may be generated by an ICT mechanism in which the charged species being transferred is proton. In the latter case, we expected the rate of the reverse reaction to be profoundly influenced by variations of the pH in the surrounding aqueous phase. However, the literature indicated that the two components of the ERP are pH independent in the physiological range of pH 5–8 [35] (see discussions in Ref. [65]). The situation with the ERP-like signal elicited from reconstituted bR is essentially the same [66]. We thus encountered an apparent paradox.

In bR membranes, interfacial proton transfer is, for obvious reason, an obligatory process. The absence of a pH-dependent photoelectric signal was therefore equally puzzling. However, from the point of view of equivalent circuit analysis, the apparent paradox was more expected than puzzling.

In other words, there is no real paradox. The apparent paradox was a consequence of a hidden parameter *chemical capacitance*, which unexpectedly distorts the measured kinetics. Recall that the true kinetics could only be recovered by means of deconvolution. The distortion would be minimal if the measurement conditions were close to short circuit. We were able to achieve the latter condition by means of the tunable voltage-clamp method. Since all previous measurements of the ERP and the ERP-like signal in reconstituted bR membranes were measured under open-circuit conditions, the apparent lack of pH dependence could be readily understood and the apparent paradox could then be resolved by the following consideration. Being predominantly RC relaxation of R_m and C_m, the open-circuit data are devoid of any significant kinetic content (explained in detail in Ref. [19]).

Shown in Fig. 6 are data of a bR model membrane, measured under a near-short-circuit condition. There are two observable components in the measured signal. A faster component $B1$ appeared immediately upon light stimulation, just like the $R1$ component of the ERP. A slower component $B2$ with the opposite polarity is sensitive to temperature, just like the $R2$ component of the ERP. The apparent temperature sensitivity of the $B1$ component is the effect of overlapping kinetics of the two components (see Fig. 6 legend for explanation). However, unlike the $R2$ component of the ERP, the $B2$ component exhibits significant pH dependence, as explained earlier.

Of course, the validity of our analysis is contingent on the universality of the equivalent circuit, which was previously devised to account for the behavior of Mg-porphyrin-containing BLM specifically. Instead of performing an exhaustive number of experiments to demonstrate (but not to prove) its generality, we resorted to an inquiry toward possible molecular mechanisms of light-induced charge separation.

There is an additional complication. Whereas the Mg-porphyrin BLM model system consists of a single reversible electron transfer reaction, the bR photosignals consists of multiple AC components in addition to a DC component. The presence of two or more components of AC photocurrent posed a problem for serious quantitative tests of the equivalent circuit, as applied to bR data. The large number of adjustable parameters would make a fit between a measured signal and the equivalent circuit fortuitous or even meaningless. Therefore, the first step should be an unequivocal decomposition of the multiple component signal. The common practice in solution phase photochemistry is to decompose a transient photosignal into as many individual exponential components as required. But this practice leads to direct violation of the zero time-integral condition. In fact, each component has a nonzero time integral and is therefore not an AC photosignal at all. This consideration shows that curve fitting does not always yield meaningful components. However, if an investigator disregarded or dismissed the presence of chemical capacitance, such a pitfall seemed

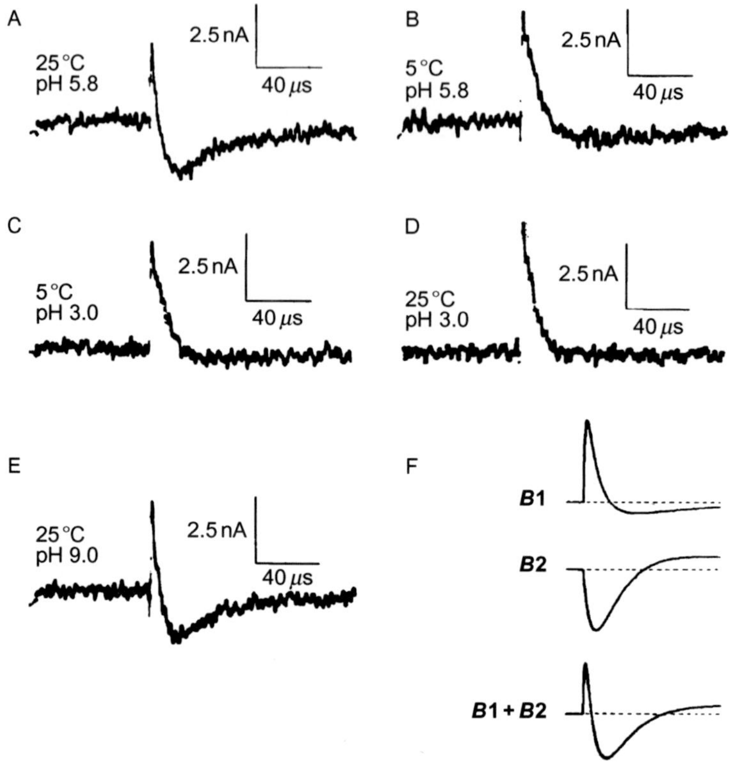

Figure 6 *Temperature and pH dependence of the photosignal from a bacteriorhodopsin thin film reconstituted by means of the method of Trissl and Montal.* The *B*1 component has a positive polarity and the *B*2 component has a negative polarity. The signals (A–E) were measured under near-short-circuit conditions. It is shown that either low temperature (5 °C) or low pH (3) reversibly diminishes the *B*2 amplitude. The concurrent increase of the *B*1 amplitude is more apparent than real. As shown schematically in (F), the *B*1 decay and the *B*2 rise have considerable overlap. Therefore, a reduction of the *B*2 amplitude leads to an apparent increase of the *B*1 peak even if the *B*1 amplitude remains constant. Adapted from Ref. [60].

nonexistent or immaterial. It took a blatant inconsistency between data reported by various laboratories to bring about the peril of ignoring chemical capacitance (see Table 1 of Ref. [67]).

5. Oriented Dipole Mechanism Versus ICT Mechanism

In validating two rival theoretical models, we avoided the unproven assumption that only one of the two models could be correct. As will be shown below, both are valid but they pertain to different components of the

AC photoelectric signal. Here, we shall interpret the equivalent circuit in terms of what is currently known about the sequence of molecular processes of light-driven proton translocation in bR membranes. The detailed mathematical derivation can be found in Ref. [65].

With the help the known molecular structure of bR and several photo-intermediates, the sequence of proton translocation is known [38–48]. Basically, there exists a proton conduction pathway consisting of a hydrogen-bonded network and extending from the intracellular surface to the extracellular surface. Proton translocation starts with light-induced isomerization of the chromophore (attached covalently to Lys-216 via a Schiff base linkage) from *all-trans* configuration to 13-*cis*. Photoisomerization causes the retinal Schiff base to suddenly lower its pK_a and to initiate deprotonation. The primary acceptor of the Schiff base proton is Asp-85. The remaining reactions of the photocycle are thermally driven. The first half of the photocycle is dominated by the transfer of a proton from the retinal Schiff base, first to Asp-85 and eventually to the extracellular aqueous phase. The active site of this sequential transfer is the highly polarized bound water 402, which is hydrogen-bonded to the protonated retinal Schiff base and two anionic aspartates, Asp-212 and Asp-85. The release of proton from the deeply buried Asp-85 to the extracellular space requires the participation of Arg-82, Tyr-57, Glu-194, and Glu-204 which, together with six bound water molecules, form a column of hydrogen-bonded networks. The second half of the photocycle is featured by reprotonation of the Schiff base from Asp-96, which, in turn, accepts a proton from the cytoplasmic space via another hydrogen-bonded chain with four bound water molecules in a single file. Kinetically, the action of the protonation switch, from an extracellular access to an intracellular access, takes place during the $M1 \rightarrow M2$ step. The $M1 \rightarrow M2$ transition incurs a major conformational change and a large entropic change; the probability of charge recombination is greatly diminished by virtue of this protonation switch. This transition is unidirectional under some conditions but reversible under others.

Schematically, the light-induced proton translocation can be cast in terms of a light-induced *coupled consecutive proton transfer reactions*, as shown in Fig. 7A [68]. Consecutive proton transfer reactions are coupled in the sense that the reactants of a reaction are actually the products of the preceding reaction and the products become the reactants of the reaction immediately following it. A_3H and A_3^- are the protonated and deprotonated Schiff base, respectively. Only five proton-binding sites are shown in the diagram for simplicity. A_1 and A_5 are the proton uptake and release sites at the cytoplasmic and extracellular surfaces, respectively. It is similar to the scheme of coupled consecutive electron transfer reactions found in the photosynthetic reaction center of *Rhodopseudomonas viridis*. The scheme shown in Fig. 7B was constructed in accordance with the known crystal structure of the reaction center [69]. A similar coupled consecutive electron

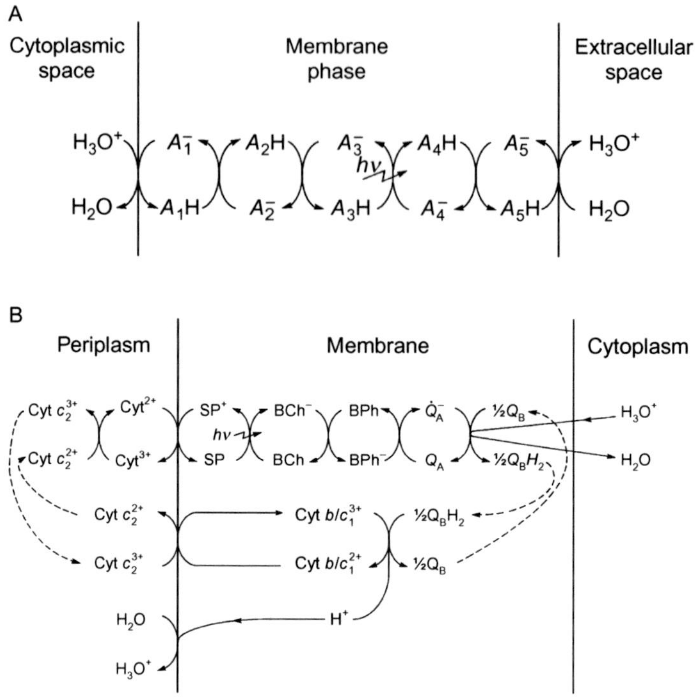

Figure 7 *Coupled consecutive proton transfer reactions along a proton transport chain in the purple membrane (A) and coupled consecutive electron transfer reactions along an electron transport chain in the reaction center of Rhodospseudomonas viridis (B).* (A) The actual number of binding sites is not precisely known. For simplicity, only five binding sites (A_1 through A_5) are shown. Site A_3 is proton-binding site of the Schiff base linkage between retinal and Lys-216. Two adjacent sites, A_2 and A_4 are Asp-96 and Asp-85, respectively. Additional sites exist between sites A_4 and A_5, and also possibly between A_1 and A_2. It is understood that the Schiff base is neutral when unprotonated, and is positively charged when protonated. Reverse proton transfer reactions are not shown. Adapted from Ref. [68]. (B) SP, "special pair" (bacteriochlorophyll dimer); *BChl*, monomer bacteriochlorophyll; *BPhe*, bacterio-pheophytin; Q_A, the tightly bound quinone; and Q_B, the loosely bound quinone. The electron transfer reactions in the cytochrome b–c_1 complex are simplified to become a single step; so are those in the four bound cytochromes on the periplasmic surface. Cyt c_2 is the peripheral protein cytochrome c_2. The dotted arrows indicate the diffusion of Q_B and cyt c_2. The concurrent proton–electron transfers occur when quinones are converted to hydroquinones, and vice versa. The involvement of quinines converts a basically transmembrane electron gradient into a transmembrane proton gradient. The converted energy is stored as a transmembrane proton gradient. Adapted from Ref. [49].

transfer reaction scheme can be constructed using detailed structural and functional information accumulated over several decades in green plant photosynthesis. A simplified and highly readable summary of green plant photosynthesis can be found in Ref. [70].

It is to be noted that the scheme in Fig. 2 represents perhaps the simplest of coupled consecutive electron transfer reactions in membranes. An important and well-established fact regarding the latter scheme is the presence of a reverse reaction (charge recombination) for each forward electron transfer (charge separation). It is therefore natural to include a reverse charge transfer reaction for each forward charge transfer reaction, including the two interfacial reactions, in the schemes of Fig. 7A and B. Figuratively, the scheme looks like the electric current flows in a cascade of RC circuits, in which the AC currents resemble the reverse electron transfers. In reality, this analogy is more than just figurative, because its equivalent circuit turned out to be a cascade of RC circuits, in which the capacitative elements are nothing but the chemical capacitance.

Thus, two types of charge separation must be considered (Fig. 8). For example, the forward proton transfer from site A_3 to site A_4 results in the formation of a transient array of electric dipoles, which disappears upon charge recombination. This process is tantamount to charging and discharging of a capacitance and the process thus manifests as an AC photoelectric signal. This process is depicted in Fig. 8 as the oriented dipole (OD) mechanism of AC photocurrent generation.

The interfacial proton transfer constitutes another kind of charge separation because binding a proton at the cytoplasmic side leaves behind a counterion in the adjacent aqueous phase (the diffuse double layer). The charge recombination is the reverse interfacial proton transfer by means of which the bound proton is released into the same aqueous phase, resulting in no net proton transport. This is similar to the interfacial electron transfer mechanism mentioned earlier and can be generalized as the ICT *mechanism* of AC photocurrent generation. It is obvious that the interfacial proton release at the extracellular surface along with its reverse reaction will lead to the generation of yet another component of AC photocurrent of ICT origin, the hypothetical $B2'$ component.

The simple model depicted in Fig. 2 applies to both the Mg-porphyrin-containing BLM and the reconstituted bR membranes. For the former system, the species A^+ shown in Fig. 8 is the electron acceptor, and A is the corresponding electron donor. For the latter system, A^+ is the proton donor, hydronium ion, H_3O^+, and A is the corresponding proton acceptor, water. Illumination with a light pulse of sufficiently short duration (about 1 μs) generates a transient sheet of space charge at the left surface of the membrane and a corresponding surface potential there. This sheet of surface charges will polarize both the adjacent (left) aqueous double layer and the opposite (right) interface, since the membrane is sufficiently thin. It is the polarization of the opposite aqueous phase that renders the transient AC photoelectric signal directly measurable.

We further noted that there is a contrast between the membrane phase and the aqueous phase. The ionic cloud relaxation time in the aqueous phase is in

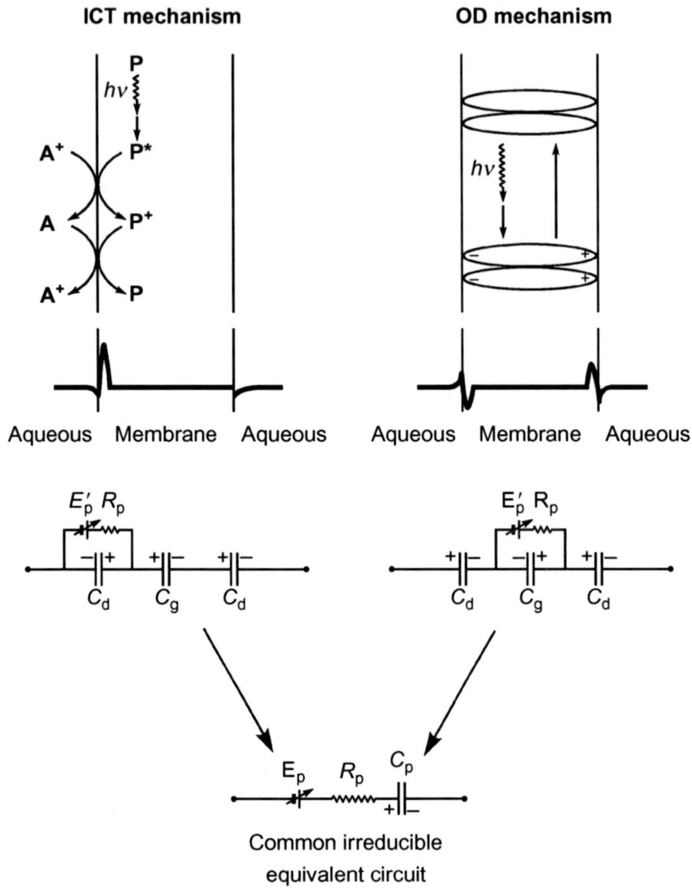

Figure 8 *Interfacial charge transfer (ICT) mechanism and oriented dipole (OD) mechanism.* The symbols P, P★, and P^+ have the same meanings as in Fig. 2. In the ICT mechanism, the symbol A^+ stands for either proton donor or electron acceptor and A stands for proton acceptor or electron donor. The reverse reaction is also shown. The OD mechanism is self-explanatory. The space charge profiles are shown just above the two microscopic equivalent circuits. The photoemf, E'_p, with an internal resistance, R_p, is located at different places with respect to the geometric capacitance, C_g, and the two double layer capacitances, C_d. The irreducible equivalent circuit is shown below them, where the three capacitances were represented by a single capacitance, C_p. The site of action of the photoemf has no counterpart in conventional analog electric circuitry. Nor has the capacitance C_p. Adapted from Ref. [65].

the subnanosecond range, whereas it lasts seconds or longer in the membrane phase. In other words, on the time scale of photoelectric relaxation (microseconds), the aqueous phase is in *instant equilibrium* but the membrane is *far from equilibrium* (as far as charge distribution is concerned). Therefore, the electrostatic description of the Gouy–Chapman theory is applicable in the aqueous phase, whereas the constant electric field condition can be applied to the membrane phase because the space charge in the membrane is confined to the left surface and virtually fixed during the time scale of interest.

The potential profile and the space charge profile across the membrane are shown in Fig. 9A. Electrification of the left interface leads to the appearance of a positive surface potential at both interfaces. The surface charges concentrating at the left interface are compensated by net negative charges accumulating at both diffuse double layers. Interpretation of the results in terms of an electric circuit (equivalent circuit analysis), however, requires linearization of the equations that describe the potential profile and the space charge profile. This treatment leads to a microscopic equivalent

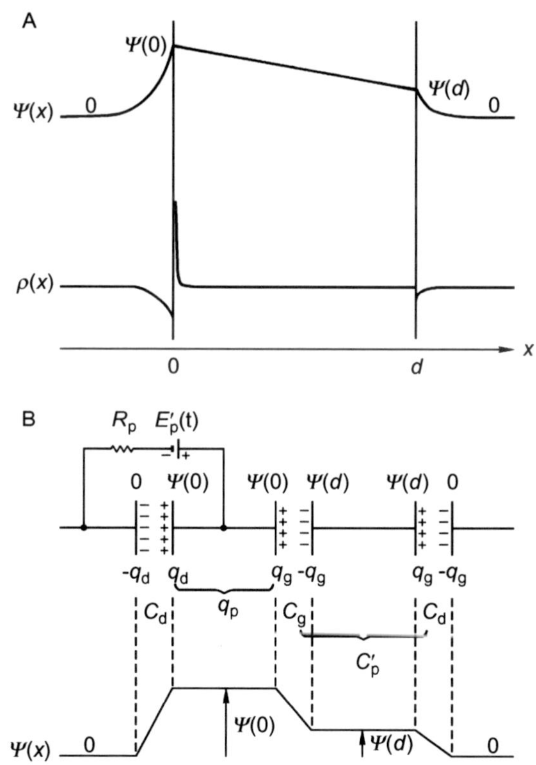

Figure 9 *Gouy–Chapman analysis of the interfacial electron transfer reaction at the oxidant interface of the system as shown in Fig.* 2. The membrane thickness is d. The potential, $\psi(x)$, and space charge distribution profiles, $\rho(x)$, are shown in (A). Linearization makes it possible to interpret the potential and charge distribution profiles in terms of an equivalent circuit shown in (B). The potential profile after linearization is also shown in (B). Adapted from Ref. [18].

circuit shown in Fig. 9B. The circuit contains two double layer capacitances, C_d, the geometric (dielectric) capacitance, C_g, and an interfacial resistance, R_p, representing the resistance encountered by the reverse charge transfer at the interface. The corresponding potential profile of this lumped circuit is also shown in Fig. 9B. The true chemical relaxation time constant (i.e., the inverse of the pseudo-first-order rate constant) is therefore R_pC_p. Note that the photoemf, E'_p, is inserted at the left interface. This circuit indicates that there are two fractions of AC photocurrent. One fraction reaches the opposite side of the membrane surface and polarizes the opposite double layer and is thus externally observable. The other fraction polarizes the adjacent double layer only and is therefore externally unobservable by conventional methods. The equivalent circuit can be further reduced to a simplified equivalent circuit, shown as the irreducible equivalent circuit in Fig. 8, in which C_g and C_d are replaced by a single capacitance, C_p, and E'_p is replaced by the effective photoemf, E_p, that is externally measurable. A resistance, R_s, is added in parallel to C_p in Fig. 2 to represent the pathway of the DC photocurrent.

The Gouy–Chapman analysis of the OD mechanism leads to a microscopic equivalent circuit, which is similar to but slightly different from that of the ICT mechanism, the main difference being the location of the photoemf, E'_p. It is natural that the photoemf in the OD mechanism be connected across the geometric capacitance, C_g, because the photocurrent source is clearly located inside the bR molecule (Fig. 8). Macroscopically, the equivalent circuit can again be reduced to one C_p connected in series with E_p. In other words, both the OD mechanism and the ICT mechanism share the same irreducible equivalent circuit (bottom part of Fig. 8).

Differentiation between the two mechanisms in relation to bR is possible on the basis of the law of mass action. An AC photosignal of ICT origin is expected to be sensitive to aqueous pH, whereas a signal of OD origin is not. As we have cautioned, an experiment designed to differentiate this pH sensitivity must be performed under conditions close to short circuit. The experiment mentioned earlier thus suggests the assignment of the *B*2 component tentatively to an ICT mechanism [60,67]. In contrast, the *B*1 component is probably generated by an OD mechanism since it seems to be independent of pH. However, a definite assignment of mechanism awaits unequivocal decomposition of the photosignal and a quantitative test of the prediction of the equivalent circuit.

With the expectation of three or more components in the bR AC photosignal, at least three different methods of measurement are needed to decompose the signal. Since exponential decomposition had been discredited, we naturally looked for another method of decomposition: physically eliminating extraneous components and isolating a pure component. In other words, we must use physical or chemical methods to separate the individual components before the isolated components could be subject to the test of equivalent circuit. In this way, we could be assured that a quantitative agreement was not fortuitous.

If bR is reconstituted into a lipid bilayer membrane, all three components, *B*1, *B*2, and *B*2′, as well as the DC component will be present concurrently. We first used a method developed by Trissl and Montal [71] to eliminate *B*2′ component and the DC component. We then developed a multilayered (ML) method [67] to isolate the pure *B*1 component. The isolate *B*1 component was demonstrated to fit the equivalent circuit (Fig. 10). Deconvolution of the isolated *B*1 signal leads to a first-order recombination time constant of 12.7 μs. The photosignal has small temperature dependence (activation energy 2.5 kcal/mol). The *B*1 component was shown to be pH independent over a wide range. As expected, these kinetic parameters are not changed at all by a change of the access impedance [67]. How we achieved the isolation of *B1* component was more clearly demonstrated by a Q-tip experiment [72].

The *B*2 component could only be obtained indirectly by subtracting *B*1 component from a composite signal, which consists of both the *B*1 and the *B*2 component. The *B*2 component was shown to have significant pH dependence and prominent temperate dependence. The existence of the *B*2′ component was demonstrated in yet another kind of reconstituted membrane, originally developed by Drachev et al. [73].

A simple picture emerges with this tentative assignment of *B*1. The photosignal is independent of aqueous pH over a wide range (pH 0–11). It is not affected by replacing water with D_2O, nor is it affected by a change of ionic strength or by treatment with fluorescamine [74]. In contrast, the composite (*B*1 + *B*2) signal obtained by the original Trissl–Montal method is sensitive to all these maneuvers. Apparently, all these sensitivities must be attributed to *B*2 alone. This dichotomous behavior between the two components suggests that we have most likely decomposed the signal in terms of discrete molecular processes. The notion that the *B*1 and the *B*2 components are natural entities is further supported by our observations on a mutant of bR D212N in which the residue 212 is changed from aspartic acid to asparagine by means of site-directed mutagenesis [75]. While the *B*1 component is apparently intact in a Trissl–Montal film reconstituted from D212N, the expected *B*2 component is missing in the range of pH from about 5 to 11.

One of the virtues of our equivalent circuit is generality [19]. The analysis of the OD mechanism and the ICT mechanism implies that the charged species being moved need not be restricted to a specific kind of charges. It is applicable whether the charges being moved are electrons, protons, or other ions. It is applicable in both natural membranes and artificial membranes. It is applicable to both visual and photosynthetic membranes. Alternatively, an *ad hoc* model must be concocted for each case, and general principles could hardly be derived from these studies.

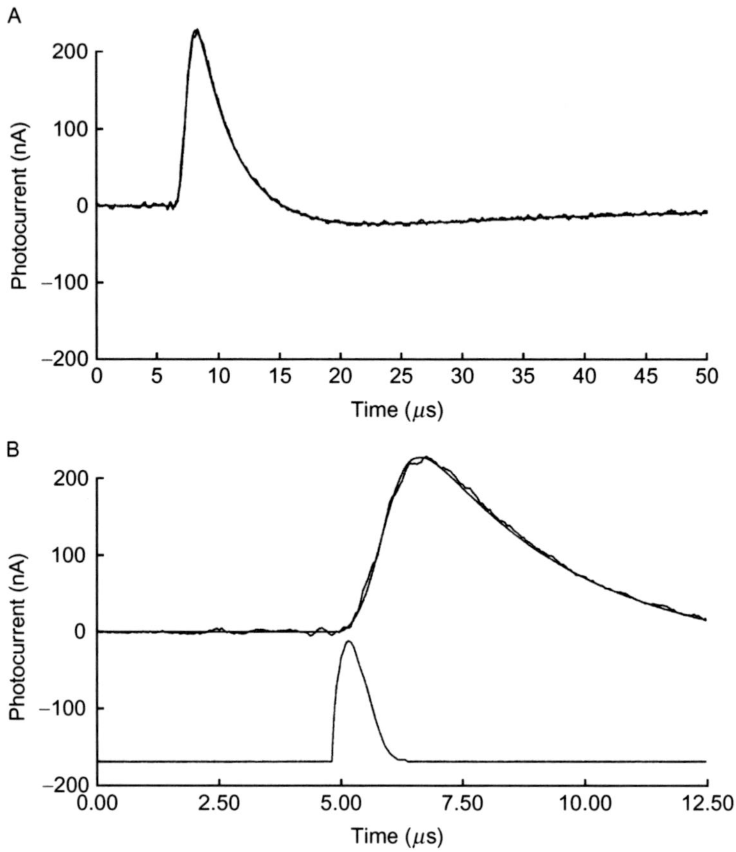

Figure 10 *Equivalent circuit analysis of the B1 component.* The ML method was used to reconstitute bacteriorhodopsin. The temperature was 25 °C. The bathing electrolyte solutions contained 0.1 M KCl and 10 mM L-histidine buffered at pH 2. The measurement was made at an access impedance of 39.2 kΩ and an instrumental time constant of 0.355 μs. The light source was a pulsed dye laser (590 nm). The light pulse was simulated with a triangular waveform with a half-height duration of 0.8 μs. The experimentally determined input parameters are $\tau_s = 3.4$ μs, $\tau_l = 24.0$ μs, $\tau_m = 6.42$ μs. $R_m = R_s = \infty$. These parameters together with R_e provide sufficient information to compute the following parameters: $R_p = 60.3$ kΩ, $C_p = 211$ pF and $\tau_p = 12.7$ μs. The computed curves (smooth traces) are superimposed on the measured curves (noisy traces) after normalization with respect to the positive peak. Normalization yields the peak photoemf of 155 mV. The same data are shown in two different scales (A and B). The measured laser pulse is also shown in (B) (bottom trace). Adapted from Ref. [67].

In brief, *ad hoc* models lack parsimony. In exchange of the additional complexity of including chemical capacitance in the equivalent circuit model, overall parsimony is achieved.

Specifically, the generality was demonstrated by the prediction of similar photosignal components in reconstituted halorhodopsin membrane. Halorhodopsin is yet another protein from *H. salinarum*. It is a light-driven chloride pump. The detail is documented elsewhere [76].

Compared to other *ad hoc* models, the generality of the concept of chemical capacitance and the associated equivalent circuit is demonstrated by our explanation of what is known as *differential responsivity* in the molecular sensor literature [77,78]. This property refers to a transient spike-like response when a pigment-containing membrane is illuminated with a long square-wave light pulse: the waveform of the recorded photosignal exhibits a positive spike upon the onset of illumination and a negative spike upon the termination of illumination. Both Koyama et al. [79] and we interpreted differential responsivity as the manifestation of light-induced charge displacement. This waveform is what is expected in a high-pass linear *RC* filter response to a square-wave voltage pulse (see Fig. 4). The capacitative reactance is initially low upon illumination and therefore the photocurrent is high. As time goes on, the chemical capacitance continues to be charged and the reactance turns high. As a consequence, the current level declines to a steady-state level if there is a DC current passage, or to the base line, if there is no such DC current (i.e., capacitatively coupled or AC coupled). However, the interpretation has remained controversial for almost a quarter century. In fact, Drachev et al. [28] proposed a drastically different interpretation and Robertson and Lukashev [29] proposed yet a third but distinct interpretation. How these latter two interpretations could be eliminated was documented in a review article [80].

Our interpretation based on the concept of chemical capacitance is, however, not perfect: the equivalent circuit of Fig. 2 cannot account for asymmetry of the two spikes. An explanation of the asymmetry will be deferred to Section 6.5.3.

6. DC Photoelectric Effect

The primary physiological function of photosynthetic membranes is to generate a transmembrane proton gradient. The transmembrane proton gradient is the first stable form of converted photon energy. This gradient is established initially by means of light-induced electron transfers across the thylakoid membrane (green plants and cyanobacteria) or across the periplasmic membrane (purple phototrophic bacteria, such as *R. viridis*). Through the intermediary action of quinoid compounds, the electron transport eventually leads to an equivalent proton transport in the opposite direction. In the purple membrane of *H. salinarum*, the generation of a proton gradient is accomplished by directly pumping protons across the purple membrane.

In either case, the efficiency and the effectiveness of the photoconversion are evaluated by analyzing the DC photoelectric effect; the AC photoelectric effect, being the manifestation of reverse reactions, is essentially an inevitable side effect that diminishes the efficiency of photoconversion. Suppression of the AC photoelectric response and/or enhancement of the DC photoelectric response can be construed as major objectives in biomembrane-based photovoltaic solar energy conversion research. By increasing the proportion of separated charges that do not recombine relative to those that do recombine, the conversion efficiency can be increased. The methodology for evaluating the DC conductivity is therefore of great technical and technological importance.

Numerous reports about the DC photoelectric effect of reconstituted membranes containing either natural or synthetic pigments appeared in the past four decades since Tien had first succeeded in eliciting photoelectric signals in such membranes. Drachev et al. [81] were the first to report the DC photoelectric effect in a reconstituted bR-containing membrane. The DC photoelectric effect was routinely investigated by using a long square-wave light pulse as the stimulating light source. Investigators routinely reported the DC photoresponses as photovoltages (under open-circuit conditions) and/or photocurrents (under short-circuit conditions). At a more fundamental level, the light-induced proton pumping activity can be analyzed in terms of the *photovoltaic* effect and the *photoconductive* effect. In equivalent circuit analysis, these effects can be evaluated in terms of the parameters *photoemf* (photoelectromotive force) and *photoconductance* (the terminology will be precisely defined later). Whereas a photovoltaic effect has been demonstrated in numerous cases, evidence in support of the existence of significant photoconductivity was often indirect and sometimes ambiguous except in special cases in which the photoconductive effect is not accompanied by a photovoltaic effect [82–84]. This is because a single act of measurement of either a photocurrent or a photovoltage does not allow for a photovoltaic effect and a photoconductive effect to be unequivocally distinguished; two independent measurements are required. This requirement is fulfilled by a measurement method, the *null current method,* which was previously developed by Hong and Mauzerall [85,86].

A membrane exhibits a photovoltaic effect if a transmembrane electric potential appears in response to illumination but vanishes upon cessation of illumination. In other words, illumination of the membrane generates a voltage source in the membrane that can be detected externally. Thus, the photovoltaic effect (light-induced emf generation) is measurable as a light-induced voltage across the membrane under open-circuit conditions. It can also be measured as a light-induced current traversing the membrane under short-circuit conditions. A photoconductive effect exists if a conductive pathway for electric current through the membrane is created, enhanced, or suppressed by illumination. A photovoltaic effect may also be accompanied by a photoconductive effect because the generation of a photovoltaic effect often requires the creation of a specific conductive pathway in addition to the pathway for ionic diffusion that has already existed in the absence of illumination and/or in the absence of the pigment. The ionic diffusion pathway allows for a transmembrane current to be driven by a transmembrane voltage. The diffusion can be mediated either by a specific ion carrier, a specific ion channel, or by a nonspecific leakage mechanism in a BLM.

The separation of the photoconductive pathway from the ionic diffusion pathway is essential in the data analysis. The important difference in the case of reconstituted bR-containing BLM is that the photoconductive pathway allows the passage of protons only, whereas the ionic diffusion pathway allows for voltage-driven passages of ions, which are often nonspecific unless special ion channel-forming or ion-carrier agents are also present. As we shall see, in a Mg-porphyrin-containing BLM, an electrochemical conductive pathway that allows electrons to be transported from one side of the BLM to the other side operates both during illumination and in the absence of illumination. In contrast, the proton conduction pathway in a bR-containing membrane operates only during illumination.

A pure photoconductive effect can exist without the concurrent photovoltaic effect, as can be demonstrated by the following example. Mauzerall and Finkelstein [82] investigated a BLM that was doped with iodine and iodide ions. The ionic conductance of the BLM is enhanced by the doping because of the formation of polyiodide ions, which have a higher partition coefficient in the membrane phase than monoiodide, I^-. Thus, the membrane-bound polyiodide ions can carry current across the BLM if a biasing potential is applied across the membrane. Illumination photochemically destroys polyiodide ions in the membrane, resulting in the reduction of this voltage-driven current. Thus, a *negative* photoconductivity without a concurrent photovoltaic effect was demonstrated because a photoresponse could not be demonstrated in the absence of the applied membrane potential. In fact, the ionic current vanishes in the absence of the biasing transmembrane voltage because there were no emf sources other than the applied biasing voltage in the BLM system. Light was just a tool to destroy the conducting element.

In a reconstituted bR membrane and a Mg-porphyrin-containing BLM, a photovoltaic effect clearly exists because a photovoltage can be demonstrated upon illumination *without* a biasing transmembrane voltage. However, the possibility of coexistence of both the photovoltaic and the photoconductive effects makes their separation somewhat problematic. For example, it is well documented that the photocurrent of a reconstituted bR membrane is initially dependent on the light intensity but reaches saturation at a higher intensity. A number of investigators have attributed this light dependence to the photoconductance, whereas we reached a different conclusion (see later). The null current method provides a general approach designed to deal with this problem in a systematic and unequivocal way.

6.1. Equivalent Circuit for DC Photoelectric Effect

The first step toward the application of the null current method is to choose an appropriate equivalent circuit. Shown in Fig. 11 are three equivalent circuits, two of which were proposed in our previous study of the Mg-porphyrin-containing BLM system [85]. Model A is consistent with the equivalent circuit for the AC photoelectric effect (Fig. 2). The chemical capacitance is omitted because no capacitative events are being considered in the DC photoelectric effect. The membrane capacitance is, however, retained for the convenience of describing the stored charges, which accompany the appearance of photovoltages. Note that the resistances R_p and R_s are combined into a single element, G_p, which is the DC photoconductance and which is equal to the reciprocal of $(R_p + R_s)$. Here, a potential source of confusion in terminology exists, and a word of clarification is warranted. The use of G_p to represent the DC photoconductance implies that it is the reciprocal of R_p. In reality, the parameter $(R_p + R_s)$ is almost identical to R_s since $R_s \gg R_p$. Thus, G_p is practically the reciprocal of R_s. This potential source of confusion is unfortunate but it can be avoided with this reminder.

Although Model A is preferred because it is consistent with the equivalent circuit for the AC photoelectric effect, Models B and C are also shown for the purpose of discussion. The differences among the three models are as follows (see Table 1). E_p in Model A drives a photocurrent through a specific proton pathway G_p but not through G_m, which represents the nonspecific ionic diffusion pathway. In contract, both E_p in Model B and E_p in Model C drive a photocurrent through both G_p and G_m. Under

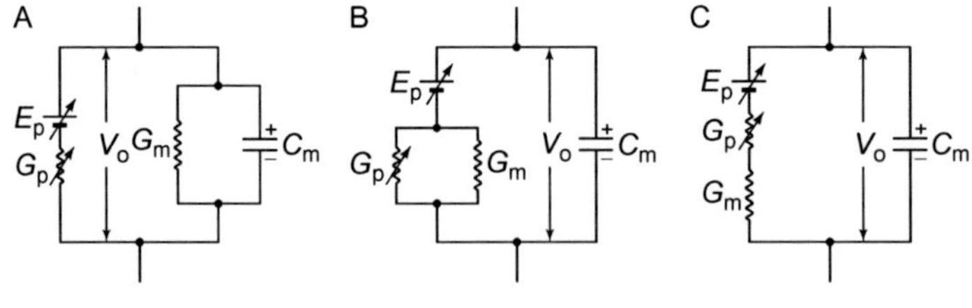

Figure 11 *Three models of equivalent circuit.* E_p is the photoemf. G_p is the photoconductance. G_m is the background ionic leakage conductance originally present in the planar BLM before the pigment is incorporated into the BLM. C_m is the membrane capacitance. V_0 is the open-circuit photovoltage. The difference of the three models is as follows. In Model A, the photoconduction channel and the ionic leakage channel are separate. The photoemf drives current through G_p but not through G_m directly. In Model B and Model C, the photoemf drives current through both of them directly. G_p and G_m are connected in parallel in Model B but in series in Model C. Adapted from Ref. [85].

Table 1 Effect of increasing G_m

Parameters	Model A	Model B	Model C
Photovoltage	Decrease	No change	No change[a]
Photocurrent	No change	Increase	Increase

[a] If G_m is initially zero, only an AC signal is observable. The addition of an ionophore brings about the DC signal.

open-circuit conditions, the capacitance C_m in Models B and C will be charged to the full value of E_p, and the open-circuit photovoltage V_0 will be equal to E_p with the top plate of C_m being positively charged, as indicated. In Model A, the capacitance C_m will be charged with the same polarity as in Models B and C but the photovoltage V_0 may or may not attain its expected full value of E_p. This is because G_m in Model A acts as a current shunt. The photovoltage V_0 attains its expected full value of E_p in Model A only if $G_m \ll G_p$. Thus, in general, the photovoltage will be diminished in Model A, but not in Model B and Model C, if G_m is increased by the action of an ionophore. However, the (short circuit) photocurrent in Model A will not be affected by increasing G_m but the photocurrent will be increased in Model B and Model C if G_m is increased by ionophore action.

Experimental tests using nystatin (an ionophore which creates a Cl^- channel in BLM) to increase the ionic conductance G_m of a Mg-porphyrin BLM revealed that the photocurrent remains constant when G_m is increased by several orders of magnitude [85]. Therefore, Model B and Model C can be ruled out. Model C has been proposed for a certain type of reconstituted membrane. Model C predicts that a plain BLM has a higher conductance than when the BLM is sandwiched with a sheet of purple membrane because G_m and G_p are connected in series in the latter. Again, this prediction contradicts experimental facts.

6.2. Null Current Method

The null current method was proposed on the basis of the principle of potentiometry in electrochemistry. The accurate value of an unknown emf source is determined under a precise condition that allows no current to flow through the emf. The accuracy is compromised if a current flow causes an error by generating a voltage drop across the internal resistance. The undesirable internal voltage drop can be avoided during a measurement by counterbalancing the unknown emf with a known and adjustable one so that no current flows through the unknown emf. The null current condition is achieved when and only when the two emf sources are equal but are of opposite polarity.

We used a voltage-clamp amplifier to implement the null current method. The command voltage E allows us to impose a transmembrane potential across the membrane. It is, therefore, also referred to as the clamping voltage. The parameters, such as photocurrent, photoemf, and photoconductance, are then measured as a function of the membrane potential, since the latter is most likely to be the independent variable that affects the rest of electric parameters. In addition, the command voltage E can also be used as the voltage source to supply an offset voltage for counterbalancing the photoemf E_p, whereas the end point of such a maneuver can be monitored via the same current output of the amplifier that reports the measured value of the photocurrent. The procedure of a null current measurement is illustrated by the schematic diagram in Fig. 12A, and an actual example is shown in Fig. 12B.

Let us consider Model A in Fig. 11. For the sake of simplicity, let us assume that G_p is much greater than G_m so that G_m can be regarded as effectively zero; that is, the parallel current path through G_m can virtually be ignored. We further assume that the conductance G_p has the same magnitude in the dark as during illumination. These restrictions will be subsequently relaxed after an intuitive picture is presented.

At the beginning of a null current measurement, the clamping voltage is set to 0 and the illuminating light source is also turned off (current level 1). While the light source remains off, the clamping voltage is then set to a preselected value V_c , and the current changes by settling at a new value (level 2) after a brief capacitative transient (not shown). While the clamping voltage is maintained at V_c, the illuminating light is then turned on. The current changes, again after a brief capacitative transient (shown in Fig. 12B but not in Fig. 12), by settling at yet another new value (level 3). The difference, level 3 minus level 2, gives the *DC photocurrent*, I_p (corresponding to the clamping voltage V_c). While the illumination continues at a fixed level, the clamping voltage is then adjusted to a new value V_c' so that the measured current is *tuned* back to level 2; that is, the endpoint of tuning is the new voltage V_c' that is required to cancel (*nullify*) the DC photocurrent. In other words, by applying an offset voltage, $-E_0 = (V_c' - V_c)$, it is sufficient to abolish the DC photocurrent and move the measured current from level 3 back to level 2 (the preillumination level).

By invoking the principle of potentiometry, the offset voltage $(V_c' - V_c)$ can be taken as equal to the photoemf (E_p) at the clamping voltage V_c, but with an opposite polarity, that is,

$$E_p = -(V_c' - V_c) = E_0. \tag{12}$$

Subsequently, the clamping voltage is then brought back to V_c, so as to ascertain that the measured current actually returns to level 3, which is the level attained before applying the offset voltage. When the light is finally

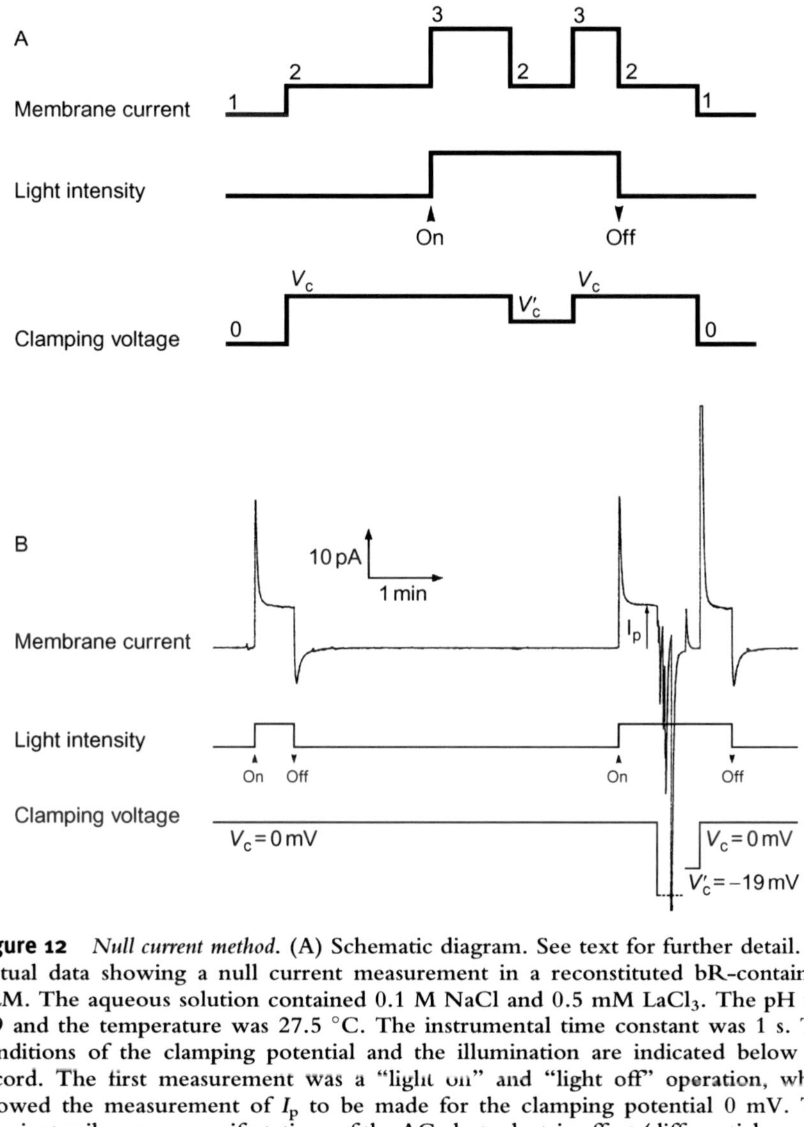

Figure 12 *Null current method.* (A) Schematic diagram. See text for further detail. (B) Actual data showing a null current measurement in a reconstituted bR-containing BLM. The aqueous solution contained 0.1 M NaCl and 0.5 mM $LaCl_3$. The pH was 6.9 and the temperature was 27.5 °C. The instrumental time constant was 1 s. The conditions of the clamping potential and the illumination are indicated below the record. The first measurement was a "light on" and "light off" operation, which allowed the measurement of I_p to be made for the clamping potential 0 mV. The transient spikes were manifestations of the AC photoelectric effect (differential responsivity). In the second measurement, the clamping potential was adjusted to obtain a null current condition, which was reached at − 19 mV. Adapted from Ref. [86].

turned off, the measured current settles back to level 2, again after a brief capacitative transient. Finally, with the light remaining off, the clamping voltage is returned to zero, and the measured current is brought back to level 1, again after a capacitative transient. This completes the cycle of measurement, in which two acts of measurement have been performed:

the measurement of the photocurrent I_p (level 3 minus level 2), and the measurement of the photoemf E_0. The DC photoconductance, G_p, is then obtained by virtue of Ohm's law:

$$G_p = \frac{I_p}{E_0}. \tag{13}$$

By varying the value V_c in a systematic way, both the DC photocurrent and the DC photoemf can be measured as a function of the transmembrane (clamping) voltage, V_c. The DC photoconductance G_p can be determined as a function of the transmembrane voltage by repeatedly using Eq. (13).

Note that the background ionic diffusion conductance in the absence of illumination, G_m, can also be obtained from the measured currents shown in Fig. 12A by means of dividing the current (level 2 minus level 1) by the applied voltage V_c, again by virtue of Ohm's law.

A null current measurement can be interpreted in an intuitive way as follows. When a net voltage of $(V' - V_c)$ is applied to the membrane during a steady illumination, the current responds by shifting from level 3 to level 2. The conductance under this circumstance (the photoconductance G_p) or, more rigorously, the combined conductance $(G_p + G_m)$ (if G_m is too large to be ignored) can be obtained in a way similar to the previous measurement of G_m. Equation (13) is thus rederived without resorting to the concept of potentiometry but simply by invoking Ohm's law instead. Thus, the null current method is similar to the method described for the measurement of G_m, except for the fact that it is being performed during illumination. In brief, G_m and G_p, as determined by the null current method, are the DC conductance measured in the absence of illumination and the DC conductance measured in the presence of steady illumination, respectively.

We shall refer to the photoemf and the photoconductance, determined by means of the above-described procedure, as the *apparent photoemf* and the *apparent photoconductance*, respectively. The adjective "apparent" implies that additional corrections of these directly measured values are required when the restrictions imposed earlier are relaxed. First, when G_m is comparable to G_p, the shunting effect of G_m cannot be ignored and it must be taken into account. Shunting diminishes the measured value of the apparent photoemf by a factor of $G_p/(G_p + G_m)$, as is intuitively evident from the consideration of the effect of a voltage divider in elementary electronics. Again by intuitive reasoning, the apparent photoconductance so measured actually contains a contribution from G_m; that is, the apparent photoconductance represents the combined value of $(G_p + G_m)$. Rigorous proof for these conclusions can be found in Ref. [86].

Another correction is required if G_p is zero in the dark so that no current can be driven through G_p by the applied potential V_c. As will be shown later for bR-containing membranes, G_p is zero in the dark and is activated by illumination to a fixed nonzero value. We shall refer to this behavior as *photogating* of the proton conduction channel. In this case, when G_p suddenly turns nonzero upon illumination, both V_c and E_p can each drive a current through G_p. Both currents will be treated as photocurrents by the definition given above.

Thus, Eqs. (12) and (13) must be replaced by

$$E_p = E_0 \cdot \frac{G_p + G_m}{G_p} - V_c \tag{14}$$

and

$$G_p = \frac{I_p}{E_0} - G_m, \tag{15}$$

respectively. In other words, the measured photoemf by means of the null current method as described by Eq. (12) is the apparent photoemf, the effectiveness of which is diminished by shunting. Furthermore, the applied voltage V_c, which drives a (voltage dependent) current through the proton conduction channel, has been misconstrued as part of the photoemf. V_c must be deducted from the apparent photoemf so as to obtain the true photoemf. In addition, the measured conductance during illumination actually contains a contribution from the dark ionic conductance. The photoconductance as determined by Eq. (13) is more appropriately referred to as the combined conductance during illumination ($G_p + G_m$), or simply apparent photoconductance for brevity. Since G_m can be independently measured, the true photoemf E_p and the true photoconductance G_p can be calculated by virtue of Eqs. (14) and (15). The rigorous derivation can be found in [86].

If G_p has the same nonzero value in the dark as well as during illumination, then the V_c driven current through G_p has already been included in the dark current. The photocurrent contains a pure component driven solely by photochemical reaction. This is the case of Mg-porphyrin-containing BLM system. The term V_c in Eq. (14) must be set to zero. Again, the rigorous derivation can be found in Ref. [86]. Thus,

$$E_p = E_0 \cdot \frac{G_p + G_m}{G_p} \tag{16}$$

and

$$G_p = \frac{I_p}{E_0} - G_m. \tag{17}$$

6.3. Null Current Analysis of DC Photoelectric Data from Mg-Porphyrin-Containing BLM

If a nonpigmented BLM is used to separate an aqueous solution of electron acceptor (e.g., 10 mM potassium ferricyanide) and an aqueous solution of electron donor (e.g., 10 mM potassium ferrocyanide), no open-circuit dark or photovoltage can be observed. If a Mg-porphyrin-containing BLM is used to separate two symmetrical aqueous salt solution without added redox reagents, no dark voltage is observed, either. As soon as redox reagents are added asymmetrically to the two aqueous phases, a dark emf appears. This emf reflects the electrochemical equilibrium of the ferricyanide/ferrocyanide couple between the two aqueous phases in the dark. Illumination increases this emf. The polarity is consistent with the asymmetry of the redox gradient and is unrelated to the direction of illumination (either from the oxidant side toward the reductant side or the opposite direction) [85].

Thus, neither the pigment alone nor the aqueous redox gradient alone generates the DC photoeffect. If 10 mM potassium ferricyanide is added to one aqueous phase and 10 mM potassium ferrocyanide to the other aqueous phase, an open-circuit voltage of about 50 mV develops in the dark within a minute after the aqueous solutions are mixed by stirring. Measurements of the dark DC conductance indicated that it is increased from that of nonpigmented BLM by a factor of 2 in a BLM that contains Mg-mesoporphyrin amyl ester (MgMAE) and by a factor of 20 in a BLM-containing MgOEP [85]. Since MgOEP is smaller in size than MgMAE, the difference suggested that diffusion of the pigment molecules is involved in the net transfer of electrons across the pigment-containing BLM.

Thus, dark redox reactions cause an increase of overall conductance across the BLM (i.e., G_p is not zero in the dark). If so, does illumination cause G_p to increase further? Experimental data show that the photoconductance of the BLM is small. The photoconductance can be measured unambiguously in the absence of a photoemf. The photoemf can be eliminated by making the redox compositions in the two aqueous phases equal because of the absence of a preferential direction for net electron transfer under such a condition. Thus, any change of the measured current upon illumination at a nonzero (imposed) membrane potential (i.e., voltage clamped at a nonzero potential) is solely attributable to the photoconductance. The results show that the DC pigment conductance deviates from the DC dark conductance with an error of less than 10% at a clamping voltage of ± 100 mV. For all practical purposes, the DC pigment conductance during illumination can be taken to be the same as its dark value (i.e., the photoconductance is negligible compared to the dark pigment conductance). Thus, G_p is practically light-independent.

The DC photoelectric signals under an asymmetric redox gradient were subjected to null current analysis, and the results for MgOEP and MgMAE

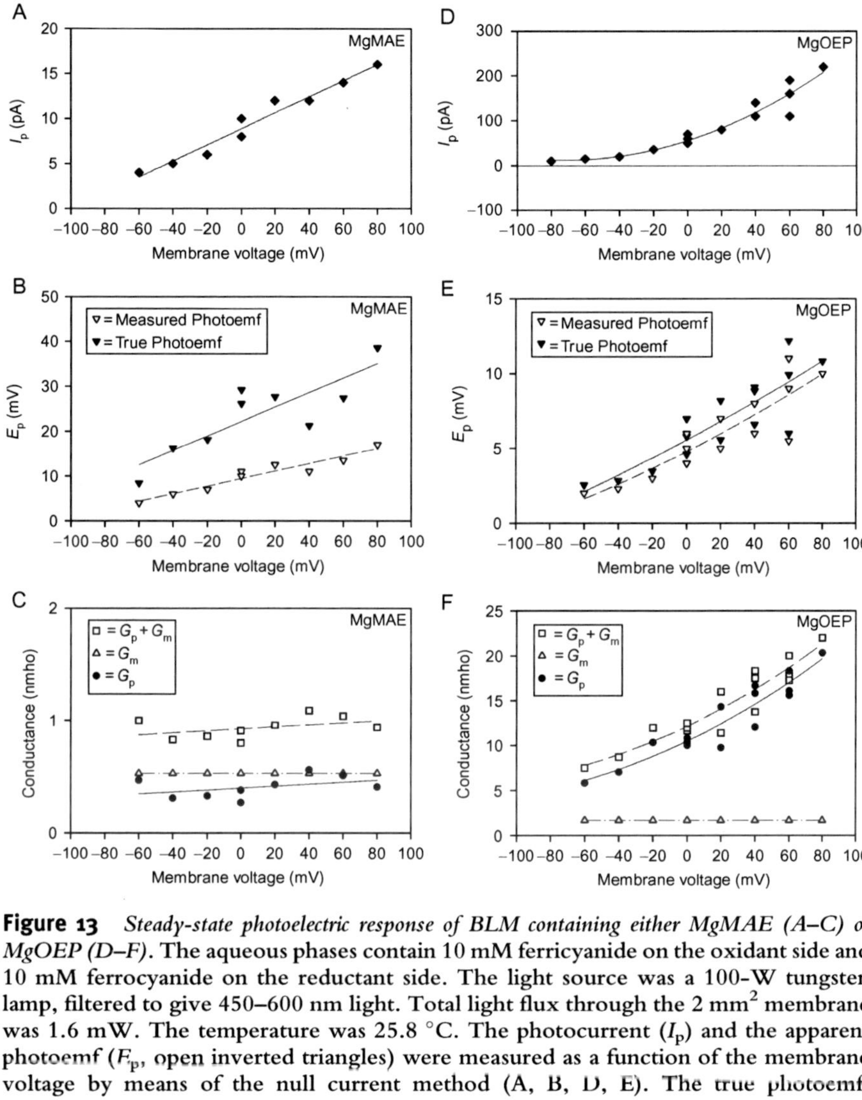

Figure 13 *Steady-state photoelectric response of BLM containing either MgMAE (A–C) or MgOEP (D–F).* The aqueous phases contain 10 mM ferricyanide on the oxidant side and 10 mM ferrocyanide on the reductant side. The light source was a 100-W tungsten lamp, filtered to give 450–600 nm light. Total light flux through the 2 mm^2 membrane was 1.6 mW. The temperature was 25.8 °C. The photocurrent (I_p) and the apparent photoemf (E_p, open inverted triangles) were measured as a function of the membrane voltage by means of the null current method (A, B, D, E). The true photoemf, calculated by means of Eq. (16), is shown as filled inverted triangles. (C and F) The apparent photoconductance ($G_p + G_m$) (open squares), which is equal to I_p/E_0, the measured dark conductance (G_m) (open triangles), and the true photoconducance (G_p) (filled circles), calculated by means of Eq. (17), are shown. All data were taken from the same membranes. The spread of points represents repeated measurements at a given voltage. Adapted from Ref. [23].

are shown in Fig. 13. It is to be pointed out that both the DC photoemf and the DC pigment conductance increase with an increasing electric field in the direction that enhances the migration of P$^+$ in the forward direction (i.e., in the direction that results in the forward reaction of the photocycle). Also note that the DC pigment conductance is higher in MgOEP-containing BLM than in MgMAE-containing BLM. These findings are expected if the pigment molecules are mobile inside the BLM. Shieh and Tien [13] reported a similar dependence of photoresponses. So did Hesketh [87], using a different kind of pigment. The difference in MgOEP-containing BLM and MgMAE-containing BLM reflects the difference of molecular weights (and molecular sizes). Note that G_p of MgMAE-containing BLM is approximately equal to G_m, and neither exhibits significant voltage dependence. In contrast, G_p of MgOEP-containing BLM is considerably greater than G_m and exhibits significant voltage dependence. Again, the differences can be understood in terms of the difference in molecular sizes. In the intensity range of 5–50 mW of continuous light illumination (530–600 nm focused to an area of the membrane of about 1 mm^2), the photoemf is found to vary approximately linearly with the light intensity (data not shown), whereas the DC pigment conductance remains virtually at its dark value. The difference in G_p of MgMAE-containing BLM and MgOEP-containing BLM is reflected in the shunting effect on the measured photoemf (Fig. 13C and D). The measured photoemf in the MgMAE-containing BLM is diminished by half, whereas the measured photoemf in the MgOEP-containing BLM is approximately the same as the true photoemf. The voltage dependence of I_p in MgMAE-containing BLM reflects the voltage dependence of the true photoemf alone, whereas that in MgOEP-containing BLM reflects the voltage dependence of both the true photoemf and the pigment conductance G_p. Indeed, the curvilinear voltage dependence of I_p can be understood as the multiplicative effect of the linear voltage dependence of I_p and G_p. The overall consistency of the null current analysis indicates that Mg-porphyrin-containing BLM behaves like a simple electric circuit, and the practice of equivalent circuit analysis is fully justified.

6.4. Null Current Analysis of DC Photoelectric Data from bR-Containing Membranes

The situation in reconstituted bR membranes is quite different. As we shall see, there is virtually no dark DC pigment conductance in bR-containing BLM but there is a prominent photoconductance. Also to be kept in mind is that experimental evidence indicates that the DC photoelectric effect is caused by proton transfer through a specific conducting pathway; bR is virtually immobilized in the membrane in the transmembrane direction and bR cannot possibly act as a proton shuttle.

For reconstitution of bR membranes, we used either the original method of Dancsházy and Karvaly [88] or its variant with a collodion film formed according to the recipe of Drachev et al. [73]. These methods, in turn, are modifications of the original BLM method developed in 1962 by

Muller et al. [5]. The light source was a continuous argon ion laser beam with the main output at 515 nm (ultraviolet outputs filtered). In the case of a BLM formed by the method of Dancsházy and Karvaly, the laser beam was focused to illuminate only the thin bilayer region in the center rather than the thick Plateau–Gibbs border. In the case of collodion films, the entire membrane was illuminated. The background ionic conductance G_m was also routinely measured. It was done before and after the fusion of purple membranes, but in the absence of illumination. Additional experimental details can be found in Ref. [86].

A typical photocurrent response (under a short-circuit condition) exhibited a transient peak at the onset of a steady illumination, followed by a sustained stationary current (Fig. 12B). Upon cessation of illumination, a negative transient peak appeared before the current level returned to the previous dark level. Qualitatively, the waveform is consistent with the prediction of the equivalent circuit shown in Fig. 2; the transient peaks are manifestations of the AC photoelectric effect (Fig. 4). However, the two peaks appeared to be asymmetric; the positive peak was more prominent and decayed faster than the negative peak. Interpretation of this asymmetry will be given later. For the purpose of investigating the DC photoelectric effect, these transient peaks can be temporarily ignored.

The stationary (DC) photocurrent (I_p) as a function of the membrane voltage is shown in Fig. 14A. The sign convention of I_p is that the positive membrane current flows from the compartment where purple membrane fragments were added (*cis* side) to the compartment free of purple membrane fragments (*trans* side). The sign convention of the applied membrane voltage is such that a positive applied voltage will drive a positive current (i.e., the *trans* side is the zero voltage reference). The voltage dependence of the photocurrent is almost linear but deviation from linearity was sometimes observed. At about -70 mV, the photocurrent became zero, and the polarity was reversed below -70 mV.

The apparent photoemf as determined directly by the null current method is shown as a function of the membrane voltage in Fig. 14B (open inverted triangles). The true photoemf, which was corrected for shunting by G_m and for the applied potential according to Eq. (14), is also plotted (filled inverted triangles). It is seen that the voltage dependence of the apparent photoemf is similar to that of the photocurrent but the true photoemf has virtually no voltage dependence. The sign convention for E_p is that a positive photoemf will drive a positive photocurrent (from the *cis* side to the *trans* side).

As explained, the apparent photoconductance, which is calculated by means of Eq. (13), contains the contribution of the ionic conductance and is therefore labeled as ($G_p + G_m$) (open squares in Fig. 14C). The ionic conductance (G_m) as a function of the membrane voltage is also plotted (open triangles). Subtraction of these two sets of data allows for the

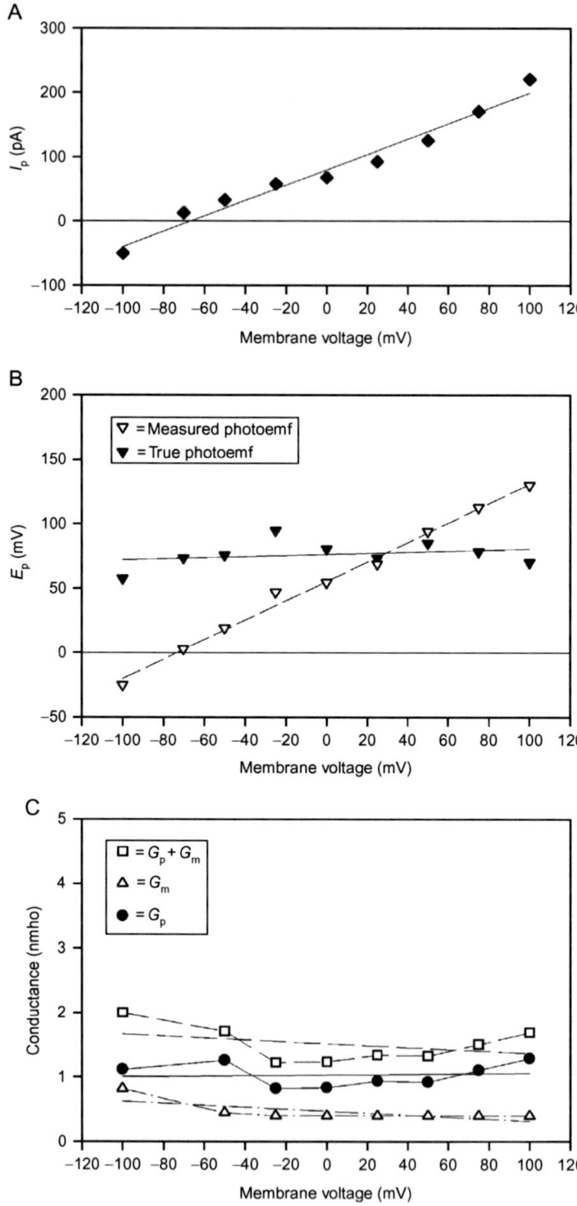

Figure 14 *Steady-state photoelectric response of a bacteriorhodopsin-containing BLM.* The aqueous solution contained 0.1 M NaCl and 0.5 mM $CeCl_3$. The pH was 6.9 and the temperature was 24 °C. (A) The measured photocurrent (I_p) as a function of membrane voltage. (B) The photoemf (E_p) as a function of membrane voltage as determined by the null current method. The apparent photoemf is shown as open inverted triangles, whereas the true photoemf, calculated by means of Eq. (14), is

calculation of the pure photoconductance G_p, which is shown in Fig. 14C as filled circles. All these conductances are ohmic (no voltage dependence).

The absolute values of conductance as well as the photocurrent and the photoemf varied from membrane to membrane, and we were thus prevented from pooling the repeated measurements together to obtain statistics. The data shown in Fig. 14 are typical. The descriptions to be presented will be based on collective features from repeated experiments under the same or similar conditions.

For a given membrane, the combined conductance during illumination ($G_p + G_m$) is consistently found to be two- to threefold higher than the dark conductance G_m. This means that G_p is either of about the same magnitude as G_m or twice as large. For conditions used in Fig. 14, ($G_p + G_m$) is 50 ± 15 nmho/cm^2, whereas G_m is 17 ± 5 nmho/cm^2. Therefore, G_p is about 33 ± 15 nmho/cm^2.

The combined conductance ($G_p + G_m$) is greater than the dark conductance. The increase is not spectacular; the photoconductance G_p is only slightly greater (about twofold) than the ionic conductance G_m. Without illumination, the incorporation of the purple membrane into the BLM does not change the background conductance (i.e., $G_p = 0$ in the dark). Illumination turns on both the photoemf E_p and the photoconductance G_p.

Unlike the claim of Bamberg et al. [89], photosignals from our reconstituted bR membranes were usually sufficiently large in amplitude for accurate measurements, without the aid of proton ionophore-induced enhancement. However, we frequently use multivalent cations to enhance the adsorption process. Occasionally, sufficiently large photosignals can be observed without the aid of multivalent cations or ionophores. It has been reported that La^{3+} has an effect on the AC photoelectric signal [73]. But we found that La^{3+} has no direct effect on the DC photoelectric signal. The presence or absence of multivalent cations does not alter the conclusions drawn for the DC photoelectric effect.

In Fig. 15, the light dependence of the photocurrent, the true photoemf, and the true photoconductance is shown. The data were collected from a bR-containing collodion membrane reconstituted according to the method of Drachev et al. [73]. The photocurrent was shown to have a linear dependence at low light intensity but eventually become saturated; any further increase of the light intensity had no additional effects (Fig. 15A). A similar light dependence is exhibited by the true photoemf (Fig. 15B).

shown as filled inverted triangles. (C) The apparent photoconductance ($G_p + G_m$) (open squares), which is equal to I_p/E_0, the measured dark conductance (G_m) (open triangles), and the true photoconductance (G_p) (filled circles), calculated by means of Eq. (15), are shown. Each set of data was fitted with a straight line by the least square method. The conductance data were also connected with line segments for the sake of easy data reading. All the data were taken from the same membrane. Adapted from Ref. [86].

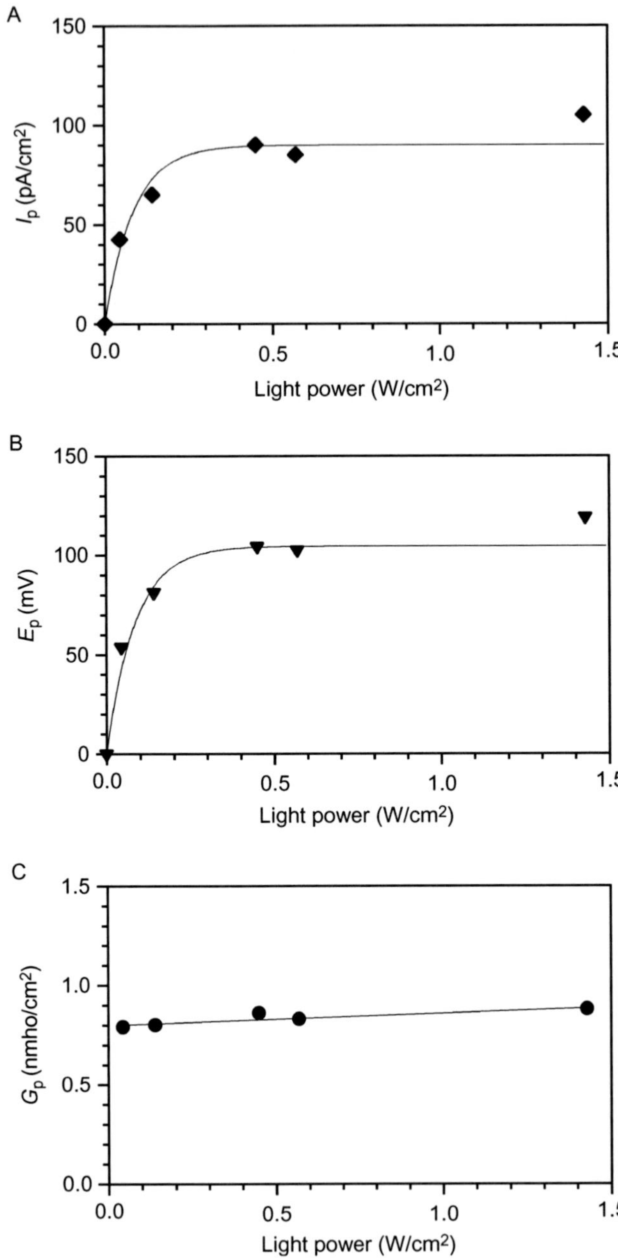

Figure 15 Light dependence of the photocurrent (A), the photoemf (B), and the photoconductance (C). Both the photoemf and the photoconductance have been corrected according to Eqs. (14) and (15). The data were from a collodion film of 2 mm^2. The ionic conductance G_m was 0.5 nmho/cm^2. The electrolyte solution contained 0.1 M KCl buffered with 50 mM L-histidine at pH 7.0. The temperature was 22 °C. Both the photocurrent and the photoemf were each fitted with an exponential curve with half-saturation light intensity of 59 mW/cm^2. The photoconductance was fitted with a straight line. Adapted from Ref. [86].

The half-saturation level of the photocurrent and the photoemf is about 0.059 W/cm^2. The true photoconductance (G_p) is shown in Fig. 15C. G_p is virtually independent of the laser light intensity (it varies $\pm 5\%$ in the range from 0.045 to 1.4 W/cm^2).

6.5. Interpretation of DC Photoelectric Data of bR-Containing Membranes

Our own data shown in Figs. 14 and 15 thus confirmed a number of observations previously reported in the literature. The photocurrent is voltage dependent and it reverses its polarity at a certain potential (i.e., the *I–V* curve is linear and intercepts the voltage axis) [81,90–94]. The photocurrent exhibits an initial linear dependence and reaches saturation as light intensity increases [95]. Beyond that, there are important differences in our interpretation. The salient feature of the null current analysis is summarized as follows.

6.5.1. Photogating of the Proton Conduction Channel

It is of interest to compare the present results of photoconductance determination in bR-containing BLM with what we observed in Mg-porphyrin-containing BLM [85]. Superficially, the magnitude of G_p shown in Fig. 14C seems to be not much different from that of G_m. However, one must realize that G_p is zero in the dark; illumination changes the overall conductance from its value in the dark (G_m) to a significantly higher value ($G_p + G_m$). Therefore, the effect of illumination is quite significant and there is little doubt that illumination opens a specific current pathway. In contrast, it is noted that G_p in Mg-porphyrin redox BLM is nonzero in the dark but illumination does not increase G_p any further [85]. This is because Mg-porphyrin-containing BLM transports electrons in the dark but bR does not transport protons in the dark (Mg-porphyrins undergo redox reactions in the dark).

The null current analysis attributed the light dependence to the photoemf rather than to the photoconductance without any prior assumptions. It is of interest to observe that the magnitude of the photoconductance is *independent of* the light intensity in the range used in this study. Of course, the photoconductance depends on light for its existence. However, once it is turned on by light, the photoconductance quickly reaches its full (saturation) magnitude in the same range of light intensity where the photoemf is still linearly dependent on light intensity. We refer to this characteristic as *photogating of the proton conduction channel* in the sense that the gate for photoconductance is opened by illumination but remains closed in the dark.

While we attribute the light dependence of the photocurrent to the same dependence of the photoemf, other investigators have assigned the dependence to the photoconductance instead. Herrmann and Rayfield [90] postulated a voltage-independent current generator shunted by a light-dependent conductance in order to interpret their data quantitatively. Szabó and Bamberg [96] reported a light-dependent photoconductance, which saturates at 0.59 W/cm^2. Here, we had made no prior assumption about light dependence in our measurements and analysis and came up with an opposite conclusion. The light-saturation level determined by Szabó and Bamberg is somewhat higher than ours. Our reported value was probably underestimated. The light intensity (power) was read directly from a light meter attached to the argon ion laser and the beam area was estimated from burning of an exposed Polaroid film by the laser beam. We therefore placed limited confidence on the absolute magnitudes of our light intensity measurements. However, the measurements of relative magnitudes of data shown in Fig. 14 are considered reliable.

6.5.2. Interpretation of Voltage Dependence of Photocurrent

By means of the null current method, we found that the apparent photoemf is linearly dependent on the applied potential but the true photoemf and the true photoconductance are not (i.e., G_p is ohmic). The voltage dependence of the photocurrent can be simply interpreted as follows. As explained earlier and documented rigorously in Ref. [86], photogating of G_p allows for the both the (true) photoemf and the applied transmembrane potential (V_c) to drive proton currents through bR's proton conduction channel. Therefore, the measured (apparent) photoemf also contains a contribution from the applied transmembrane potential. The photocurrent thus contains a constant fraction, which is driven by the true photoemf, and an additional fraction, which is driven by the applied transmembrane potential and which is linearly proportional to the applied potential. (Note that the linear voltage dependence of the latter fraction implies that the proton conduction channel is ohmic.) Incidentally, this feature is not a consequence of the shunting effect. If the V_c driven fraction of the photocurrent is excluded, the remaining fraction becomes independent of the applied potential. We thus concluded that the true photoemf is independent of the applied transmembrane potential. The polarity reversal of the photocurrent observed at -70 mV is caused by the increasing influence of an opposing applied transmembrane potential. Thus, the determination of the intercept of the voltage axis is an alternative way to determine the photoemf (which is a common practice in I–V analysis). The value 70 mV compares favorably with the average values of the photoemf as determined by the null current method, 77 ± 10 mV. The DC photoelectric property of bR thus appears to be simple with this interpretation. Ockham's razor seldom shines so brightly in biology. It is of interest to note that no such corrections are necessary for the previously described Mg-porphyrin redox BLM system [85].

Bamberg et al. [92] proposed an alternative interpretation of the polarity reversal. These investigators suggested that the polarity reversal is caused by the presence of two populations of purple membrane sheets in the system, with the majority of the purple membrane sheets being incorporated in one orientation, but a much smaller fraction being oriented in the opposite direction. Thus, when the photocurrent decreases as negative voltage is applied, the increasing photosignal attributed to the other minor fraction might be revealed. However, this explanation is inconsistent with the observation of Gavach et al. [95], which was also confirmed in our laboratory. Whereas the stationary photocurrent is reversed at a certain applied voltage, the transient positive peak is not reversed. This transient peak reflects the AC photoelectric current and its polarity can be regarded as an indication of the orientation of the majority of bR molecules inside the model membrane. Gavach et al. interpreted the voltage dependence as the effect of variation of the surface potential of the membrane, which influences the degree of protonation of certain donor sites that are separated from the site of the photoemf.

Nagel et al. [97] succeeded in expressing bR in oocytes of *Xenopus laevis* and measured the DC photocurrent directly by means of the patch-clamp method. Again, they found nearly linear voltage dependence of the DC photocurrent. However, no polarity reversal was observed up to -165 mV. They cited this lack of polarity reversal as the support for the interpretation of polarity reversal as the consequence of two populations of bR orientation, previously proposed by Bamberg et al. [92]. In their patch-clamp measurement, presumably there is only one orientation of bR in such a way that all bR molecules pump proton in the outward direction. Possible nonlinearity of the voltage dependence prevents extrapolation of the photocurrent to estimate the voltage where polarity reversal can be achieved. The presence or absence of polarity reversal can only regarded as unsettled for direct patch-clamp measurements. In a subsequent report by Nagel et al. [98], the voltage-clamp measurements show little evidence of nonlinearity in the range of -160 to $+60$ mV. Our interpretation of the voltage dependence in terms of the photogating of the proton conduction channel thus appears much simpler than these alternative interpretations.

6.5.3. Interpretation of Asymmetric Waveform of the AC Photocurrent

We are now ready to return to the discussion of differential responsivity. Elementary analysis indicates that a linear *RC* high-pass filter exhibits a waveform similar to what is evident in Fig. 12B and that the positive spike and the negative spike should be symmetrical. Analysis based on the AC photoelectric effect explains the waveform characteristic of the high-pass filter response. But the latter does not explain the slight asymmetry shown in Fig. 12B (c.f., Figure 1 in Ref. [91]). The positive spike peaks more prominently and decays considerably faster than the negative spike. This observation is consistent with the fact that G_p is zero in the dark. The hastened relaxation during illumination is likely caused by the sudden increase of G_p, which is equivalent to a light-induced decrease of the resistance designated as R_s in the equivalent circuit used previously for the AC photoelectric effect (Fig. 2; $R_s = \infty$ in the dark, but $R_s = 1/G_p$ in the light).

The asymmetry of the two spikes shown in Fig. 12B is not unique to the bR system. Shown in Fig. 16A is an open-circuit voltage measurement recorded intracellularly from an intact giant chloroplast of *Peperomia metallica* [99]. The waveform of the photoresponse to a square-wave light pulse stimulation looks similar to what we reported in Fig. 12B. Upon the addition of valinomycin (a K^+ ionophore) in the presence of 30 mM K^+, the decay of both spikes became faster than the control but retained the asymmetry of their amplitudes, whereas the DC photovoltage level became diminished (shunting effect). A similar waveform also appears in a BLM reconstituted from purified reaction center of *Rhodobacter sphaeroides* [100] (Fig. 16B) and from subchromatophore pigment–protein complexes of *Rhodospirillum rubrum* [28,101]. These observations are consistent with the concept of chemical capacitance and photogating taken together.

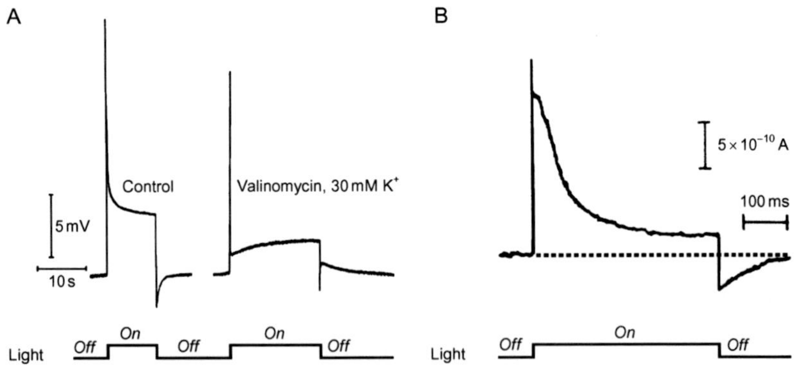

Figure 16 *DC photoelectric signals from photosynthetic membranes.* (A) The open-circuit photovoltage was recorded intracellularly from an intact chloroplast of *Peperomia metallica*. The control signal before the addition of valinomycin is shown at the left and the effect of adding 1 μM valinomycin and 30 mM potassium ions in the external medium is shown on the right. (B) The photosynthetic reaction center of *Rhodobacter sphaeroides* (8 μM) supplemented with 100 μM of ubiquinone-10 was reconstituted into a BLM. One aqueous phase contains 25 μM reduced cytochrome *c*, and the other aqueous phase contains 1 mM ferricyanide. See text for explanation. Adapted from Refs. [99] and [100].

7. Applications of DC Photoelectric Effect in Artificial Solar Energy Conversion

For a number of years since its discovery, bR has been a favorite biomaterial of investigators who devoted their efforts to the investigation of artificial solar energy conversion. That a reconstituted bR membrane or thin membrane is a photovoltaic device has been amply demonstrated by many investigators. Considerable efforts have been devoted subsequently to the description of the proton conduction pathway. Here, we shall discuss the electrical characteristics of the bR membrane that are relevant for solar energy conversion.

A bR-containing membrane or thin film has sometimes been described as a photodiode. However, our experimental result described in Section 6 indicates that the proton conduction pathway is not *rectified*: the current driven by an applied transmembrane potential encounters the same resistance in either direction during illumination; that is, bR is not a photodiode in the strict sense (Fig. 17). The notion that the photoemf and the applied transmembrane potential are interchangeable for driving a transmembrane proton current is, however, consistent with the basic principle of bioenergetics.

According to the chemiosmotic theory [102,103], the converted energy by bR photoreaction is stored as a transmembrane electrochemical gradient of protons. Furthermore, both the electrical component (ΔV) and the chemical component (ΔpH) of this gradient are equivalent and both are available for ATP (adenosine triphosphate) synthesis. Thus, our photoelectric data provide a *quantitative* demonstration that the two parts of electrochemical energy are interchangeable bioenergetically. Metaphorically, the process is tantamount to equal currency exchange rates regardless of "buying" or "selling." Were there a rectification, the interconversion would not be equal because the voltage drop would be more in the back-biasing direction than in the forward-biasing direction, which is tantamount to a difference between "buying" and "selling" rates in currency exchange, metaphorically speaking.

Rectification is an important property of a photodiode that enables it to minimize energy loss due to charge recombination; the resistance, which the photocurrent encounters during (internal) charge recombination, is much higher than the forward resistance during charge separation. As for achieving good energy conversion efficiency, bR apparently relies on a different strategy than rectification. We argue here that photogating allows bR to prevent wasteful dissipation of converted energy in the dark.

The generation of a transmembrane proton gradient is the intermediate step between photon energy conversion and bioenergetic synthesis of ATP

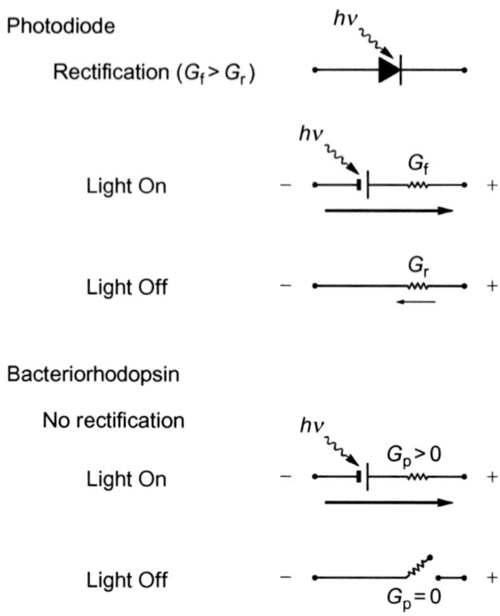

Figure 17 *Schematic diagram comparing rectification in a conventional photodiode and photogating in bacteriorhodopsin.* G_f is the forward conductance and G_r is the reverse conductance of the photodiode. Photoactivation causes a photocurrent to flow in the forward direction as indicated. As a consequence, the photodiode becomes back biased. Back biasing drives a current in the reverse direction during illumination but the reverse current encounters a much greater resistance because of rectification ($G_r \ll G_f$). In the dark, charge recombination is minimized because the conductance G_r is much smaller than G_f. In bacteriorhodopsin, there is no rectification: the forward photocurrent and the reverse current (driven by back biasing) encounter the same resistance ($1/G_p$). Thus, the observed photocurrent is actually the difference of these two currents. The constancy of the measured true photoemf indicates that illuminated bacteriorhodopsin's ability to maintain a net "driving force" is not affected by the extent of back biasing. In other words, the performance of bacteriorhodopsin during illumination has not been seriously compromised by the lack of rectification but it would be seriously compromised if the proton conduction channel remained open in the dark. Photogating of the proton conduction channel averts this potential problem: $G_p = 0$ in the dark. The plus and the minus signs indicate the direction of biasing. In the case of bacteriorhodopsin, the signs also indicate the polarity of the photovoltage. Adapted from Ref. [22].

and other energy rich compounds known as reducing equivalents, such as NaDH and NADPH. The formation of the transmembrane electrical gradient (i.e., the photovoltage) constitutes a situation similar to back biasing of a photodiode. Back biasing provides the driving force to drive protons back into the cell.[1] In principle, there are several ways the electrochemical

[1] Strictly speaking, the proton backflow is driven both by the electrical component and by the chemical component, of the electrochemical potential.

potential can drive protons back to the cytoplasm: (a) protons reentering the cell via a reverse proton flow through the same proton channel in bR (internal charge recombination); (b) protons or other small ions reentering the cell via leakage through the phospholipid portion of the purple membrane, and thus dissipating the converted energy (shunting); (c) protons reentering the cell via the proton channel of ATP synthase (residing in the red membrane), thus synthesizing ATP; (d) protons reentering the cell via the flagella motor apparatus, thus powering the motion of the flagellas; and (e) protons reentering the cell to power a cotransport system (symport or antiport). The first two routes would lead to wasteful dissipation of the converted energy. They are prevented (or minimized) in the purple membrane by photogating and by having a relatively small area of phospholipid portion in the two-dimensional crystalline structure of the purple membrane [104]. In the reconstituted bR BLM, G_p in the dark is considerably smaller than G_m. This property means that the insulation against a proton backflow via the proton conduction channel in the dark is much more effective than the insulation of the phospholipid portion of the artificial BLM. Quantitative comparison of G_p and G_m in a native purple membrane is not available. However, we think that the ratio G_p/G_m should be much greater in the native purple membrane than in our experimental system because reconstitution in our experimental system could not have achieved the same high density of bR packing as in the native purple membrane. From a mechanistic point of view, rectification in a photodiode is more important in the dark than during illumination. Whereas light-induced charge separation in a photodiode encounters the forward resistance ($1/G_f$ in Fig. 17), charge recombination during illumination encounters a much higher resistance ($1/G_r$ in Fig. 17). Nevertheless, net forward charge separation continues because it is being pumped by light. In the absence of illumination (photon pumping), without a higher resistance ($1/G_r$ in Fig. 17), the previously separated charges would recombine just as rapidly. The bR data indicate that photon energy conversion is still possible without rectification. The lack of rectification in the purple membrane may be the consequence of the universal presence of reverse reactions in the scheme of coupled consecutive charge transfer reactions (Fig. 7). However, net forward charge transfers are still possible, in spite of the lack of rectification. Thus, by means of photogating, bR achieves the same goal (in the dark, at least) as does a photodiode by means of rectification.

We further suspect that photogating of a light-dependent proton conduction channel may also be present in a chlorophyll-based photosynthetic membrane. As shown in Fig. 16A and B, the asymmetrical transient current spikes that appear upon the onset and the cessation of illumination of a giant chloroplast from *P. metallica* and of a reconstituted membrane from the reaction center of *R. sphaeroides* suggest that the charge conduction pathway in the photosynthetic reaction center may also be activated only by light.

Evidence from direct electrical measurement of G_p (as compared to G_m) is presently lacking. But if this interpretation is correct, then it appeared that Nature had implemented the same design principle by using completely different molecular constructs.

8. Concluding Summary

In this chapter, we consider the problem of biology-based solar energy conversion as an alternative source of energy. However, we restrict our discussion to photovoltaic solar energy conversion owing to our own limitations of expertise. All naturally occurring photoconverting biodevices are membrane based: photopigments are embedded asymmetrically in lipid bilayer membranes. The asymmetry is a prerequisite for the photovoltaic effect to generate useful electric energy or, more precisely, electrochemical energy.

We approach this problem by gaining engineering insights into biological systems. We also exploit the reverse approach. We used fundamental principles of man-made electric devices to formulate research strategies of biological photoconverting systems. Structurally speaking, biological systems are formidably complex. However, the technique of forming bilayer lipid membrane (BLM) *in vitro* and a number of assorted techniques inspired by the original BLM method allowed investigators to develop model systems, which were amenable to mathematical analysis and quantitative measurements. Essentially, a BLM-based model system fulfills the reductionist's dream by preserving essential features but eliminating extraneous features that had made rigorous analysis intractable because of inherent complexity. The advantage is immense and obvious. In science history, many qualitative arguments or models eventually turned out to be quantitatively wrong. BLM-based model system studies allowed investigators to eliminate fortuitous agreements between theories and experiments at an early stage.

The primary process in photobiological membranes is light-induced charge separation, in which the absorbed light energy is temporarily stored. In the case of bR, this energy is utilized to drive subsequent conformational changes, which makes possible long-distance charge separation by dividing the process into many small substeps. Eventually, charges are separated across the entire thickness of biomembranes. In this way, charge separation across the entire span of BLM is stabilized since phospholipid bilayers are good insulators that discourage charge recombination. In chlorophyll-based photosynthetic membranes, the charge separation, in the form of electron–hole generation, is further propagated along a chain of electron acceptors/donors. Again, long-distance charge separation is subdivided into smaller substeps. Again, in this way, the separated charges are stored across the membrane. Comparison of these different systems reveals a common feature. The scheme of successive charge separation can be viewed as *coupled consecutive charge transfer reactions*.

With this insight, we analyzed two model systems with various degrees of complexity. In the very outset, engineering considerations led us to realize that the first-order business was to develop effective measurement methodology. We found that direct transplantation of classical electrophysiological methodology was not the answer. However, lessons learned in classical electrophysiology served as guidance for developing relevant model systems as well as methodology. It did not take long for us to realize that the key to success was "intermarriage" of electrophysiology and electrochemistry.

In the first BLM-based model system analyzed in this article, the chemistry was kept at a bare-bone level. It allowed us to develop the relevant equivalent circuit analysis and two measurement methods, the tunable voltage-clamp and the null current method. It was a fruit of interplay of biology and engineering concepts. The separation of photoelectric effects into the AC effect and the DC effects vastly simplified the problem and spared us unnecessary confusions, not to mention wasting available resources. Superficially, only the DC photoelectric effect is relevant to solar energy conversion, whereas the AC photoelectric effect is an unwanted side effect that accompanies premature (internal) charge recombination. Nevertheless, understanding of its implication helps interpretation of relevant experiments. In hindsight, studying the AC effect was a necessary step in achieving a better understanding of photoconversion.

By means of these model systems, it became clear that photoelectric measurements were tricky matters that need to consider an unusual form of electric current injection. Since the electric generators are membrane based, electric current injection takes place inside the membrane or at the membrane–water interface, which is shielded from outside by the so-called diffuse double layer. This situation has no counterpart in conventional analog electronics or in classical electrophysiology. It is necessary to introduce the concept of chemical capacitance. Even in hindsight, a combined approach using equivalent circuit analysis and electrochemistry appeared to be the only effective way to unmask to complex phenomenology and to achieve unequivocal interpretation of the measured photoelectric data.

By combining electrochemical analysis and conventional chemical kinetic analysis, we were able to rederive the equivalent circuit, which was initially obtained empirically by analyzing the waveform of photosignals. Two prototype mechanisms of the AC photoelectric effect were identified. These mechanisms in turn suggested strategies to "dissect" a composite photosignals when we moved up to the next level of complexity: bR model membranes.

The photovoltaic effect and the photoconductive effect are two major effects of relevance to solar energy conversion. Classical electrophysiological techniques are inadequate to handle these two effects especially when both coexist. The null current method was designed with separation of these two effects in mind.

Because of the presence of a photoemf and a photoconductance, conventional electrophysiological methodology cannot determine all of the relevant circuit parameters of the DC photoelectric effect. By incorporating the principle of potentiometry, the null current method can effectively deal with the problem. In the two prototype systems being analyzed in this chapter, specific photoconduction channels, enabled by photochemical reactions, are connected in parallel to the nonspecific ionic conductance. Both the photogenerated and the voltage-driven currents pass through the photoconduction channel. However, the photoconduction channel of bR opens only during illumination, whereas that of the Mg-porphyrin BLM opens whether the membrane is illuminated or not. Therefore, the photocurrent in the bR-containing BLM consists of two fractions: a light-dependent, voltage-independent proton current and a light-independent, voltage-dependent proton current. In contrast, the voltage-driven electron current in Mg-porphyrin-containing BLM is counted (measured) as a dark current; the photocurrent is exclusively generated by photoreactions. The null current method allows for unequivocal separation of these two separate fractions of current without prior assumptions.

As far as light energy is converted by means of charge separation, the issue of unwanted and premature charge recombination is always a technical concern. Rectification is an inherent property of photodiodes that minimizes the unwanted charge recombination. However, as revealed by analysis of the bR system, rectification is neither mandatory nor the only way to minimize unwanted charge recombination. Unwanted charge recombination is not a serious drawback during illumination but it is a critical matter in the absence of illumination. bR resorts to a different mechanism: the proton conduction channel is turned off completely in the absence of illumination. Whereas this feature could be demonstrated by direct measurement and null current analysis in bR-containing BLM, we suspect that it might be a universal mechanism in other photoconverting biomembranes. We thus invoked indirect evidence made possible by a phenomenon known as differential responsivity. Essentially, it is a manifestation of the presence of chemical capacitance, although controversial interpretations existed in the literature. Differential responsivity has been exploited for construction of a motion detector [77]. The transient waveform when light-sensitive membranes are illuminated with a square-wave light pulse reflects the AC photoelectric effect. It is an inherent feature of an *RC* high-pass filter. Such a circuitry can also be used as a differentiator; the waveform appears when the level of illumination changes. The equivalent circuit in Fig. 2 predicts a symmetric waveform. But the waveform observed in bR membranes and other photosynthetic membranes often exhibited asymmetry. It turned out that the asymmetry is related to a phenomenon, which we call photogating of bR's proton conduction channel. The null current analysis reveals that the parameter R_s needs not be a constant. When it is no longer a constant, the waveform of differential responsivity may no longer be symmetrical. The asymmetry might thus betray the presence of the photogating phenomenon, which was first demonstrated in the bR membrane. This mechanism of minimizing internal charge recombination in the dark might be more universal than we had initially suspected.

Finally, a comment is in order with regard to mathematical modeling of biological phenomena. An unspoken guiding principle in science is often referred to as Ockham's razor, which dictates that modeling needs not be made unnecessarily complex. In light of Ockham's razor, let us examine our present approach with other rival approaches. Most conventional approaches analyzed kinetic data by breaking a signal into exponential components. While this was a routine approach in solution phase chemistry, it could be problematic in analyzing photoelectric data since electrical interaction between photoactive components and the inert supporting membrane can distort the signal and make it difficult to interpret. The introduction of the concept of chemical capacitance seemed to make the analysis more complex, and the approach seemed to run contrary to the advice of Ockham's razor, but the resulting simplicity resides in the universality of the equivalent circuit and parsimony. For example, the clear-cut dichotomy of light dependence of the photoemf and the photoconductance shown in Fig. 15 could hardly be achieved by means of many *ad hoc* models of limited validity. The complication brought about by chemical capacitance is just a tradeoff, which paid dividends in terms of deeper understanding. In the realm of technological applications, only through understanding can real progress be made. In anticipation of the ultimate skyrocketing rise of the oil price and the inconceivable consequence of ever elevating carbon dioxide level in the atmosphere, it is never too early to invest in basic research of solar energy conversion in Nature.

ACKNOWLEDGMENTS

The author acknowledges the contribution of the following individuals whose experimental work on bacteriorhodopsin was cited: Man Chang, Albert Duschl, Brian Fuller, Filbert Hong, Sherie Michaile, Baofu Ni, Ting Okajima; Michelle Petrak, and Wita Wojtkowski. The author wishes to thank his collaborators: Janos Lanyi, Lowell McCoy, Mauricio Montal, and Richard Needleman. The author is indebted to Prof. David Mauzerall of The Rockefeller University, under whose supervision the work on magnesium porphyrin-containing bilayer lipid membranes was performed. The author cherishes the friendship and professional support kindly extended by the late Prof. H. Ti Tien.

REFERENCES

[1] N.S. Lewis, D.G. Nocera, Powering the planet: chemical challenges in solar energy utilization, Proc. Natl. Acad. Sci. USA 103 (2006) 15729–15735.
[2] A. Regalado, Reinventing the leaf, Sci. Am. 303(4) (2010) 86–89.
[3] D. Gust, T.A. Moore, A.L. Moore, Solar fuels via artificial photosynthesis, Acc. Chem. Res. 42 (2009) 1890–1898.
[4] R. Eisenberg, Rethinking water splitting, Science 324 (2009) 44–45.
[5] P. Mueller, D.O. Rudin, H.T. Tien, W.C. Wescott, Reconstitution of cell membrane structure *in vitro* and its transformation into an excitable system, Nature 194 (1962) 979–980.
[6] P. Mueller, D.O. Rudin, Induced excitability in reconstituted cell membrane structure, J. Theor. Biol. 4 (1963) 268–280.
[7] P. Mueller, D.O. Rudin, Action potential phenomena in experimental bimolecular lipid membranes, Nature (London) 213 (1967) 603–604.
[8] P. Mueller, D.O. Rudin, Action potentials induced in bimolecular lipid membranes, Nature (London) 217 (1968) 713–719.
[9] P. Mueller, D.O. Rudin, Resting and action potentials in experimental bimolecular lipid membranes, J. Theor. Biol. 18 (1968) 222–258.
[10] H.T. Tien, Light-induced phenomena in black lipid membranes constituted from photosynthetic pigments, Nature 219 (1968) 272–274.
[11] H.T. Tien, Photoelectric effects in thin and bilayer lipid membranes in aqueous media, J. Phys. Chem. 72 (1968) 4512–4519.
[12] H.T. Tien, Photoelectric bilayer lipid membrane: a model for the thylakoid membrane, Brookhaven Symp. Biol. 28 (1976) 105–131.
[13] P.K. Shieh, H.T. Tien, Photoredox reactions in pigmented bilayer lipid membranes, Bioenergetics 6 (1974) 45–55.
[14] H.T. Tien, N.B. Joshi, Photoinitiated electron transfer processes on and across bilayer lipid membranes, Photbiochem. Photobiophys. 10 (1986) 241–251.
[15] H.T. Tien, Bilayer Lipid Membranes (BLM): Theory and Practice, Dekker, New York, 1974.
[16] H.T. Tien, A. Ottova-Leitmannova, Membrane Biophysics As Viewed From Experimental Bilayer Lipid Membranes (Planar Lipid Bilayers and Spherical Liposomes), Elsevier, Amsterdam, 2000.
[17] H.T. Tien, Membrane photobiophysics and photochemistry, Prog. Surface Sci. 30 (1989) 1–200.
[18] F.T. Hong, Charge transfer across pigmented bilayer lipid membrane and its interfaces, Photochem. Photobiol. 24 (1976) 155–189.
[19] F.T. Hong, Displacement photocurrents in pigment-containing biomembranes: artificial and natural systems, in: M. Blank (Ed.), Bioelectrochemistry: Ions, Surfaces, Membranes, American Chemical Society, Washington, DC, 1980, pp. 211–237(ACS Advances in Chemistry Ser. 188).
[20] H.-W. Trissl, Photoelectric measurements of purple membranes, Photochem. Photobiol. 51 (1990) 793–818.
[21] F.T. Hong, Electrochemical processes in membranes that contain bacteriorhodopsin, in: M. Blank, I. Vodyanoy (Eds.), Biomembrane Electrochemistry, American Chemical Society, Washington, DC, 1994, pp. 531–560(Advances in Chemistry Ser. 235).
[22] F.T. Hong, Interfacial photochemistry of retinal proteins, Prog. Surface Sci. 62 (1999) 1–237.
[23] F.T. Hong, Interfacial photochemistry in biomembranes, in: H.S. Nalwa (Ed.), Handbook of Photochemistry and Photobiology. vol. 4, American Scientific Publishers, Stevenson Ranch, CA, 2003, pp. 383–430.

[24] A.L. Hodgkin, A.F. Huxley, Currents carried by sodium and potassium ions through the membrane of the giant axon of *Loligo*, J. Physiol. 116 (1952) 449–472.
[25] A.L. Hodgkin, A.F. Huxley, A quantitative description of membrane current and its application to conduction and excitation in nerve, J. Physiol. 117 (1952) 500–544.
[26] J.-H. Fuhrhop, D. Mauzerall, The one-electron oxidation of magnesium octaethylporphin, J. Am. Chem. Soc. 90 (1968) 3875–3876.
[27] J.-H. Fuhrhop, D. Mauzerall, The one-electron oxidation of metalloporphyrins, J. Am. Chem. Soc. 91 (1969) 4174–4181.
[28] L.A. Drachev, V.N. Frolov, A.D. Kaulen, A.A. Kondrashin, V.D. Samuilov, A. Yu. Semenov, et al. Generation of electric current by chromatophores of *Rhodospirillum rubrum* and reconstitution of electrogenic function in subchromatophore pigment-protein complexes, Biochim. Biophys. Acta 440 (1976) 637–660.
[29] B. Robertson, E.P. Lukashev, Rapid pH change due to bacteriorhodopsin measured with a tin-oxide electrode, Biophys. J. 68 (1995) 1507–1517.
[30] F.T. Hong, D. Mauzerall, Photoemf at a single membrane-solution interface specific to lipid bilayers containing magnesium porphyrins, Nature New Biol. (London) 240 (1972) 154–155.
[31] F.T. Hong, T.L. Okajima, Electrical double layers in pigment-containing biomembranes, in: M. Blank (Ed.), Electrical Double Layers in Biology, Plenum Press, New York, London, 1986, pp. 129–147.
[32] F.T. Hong, D. Mauzerall, Interfacial photoreactions and chemical capacitance in lipid bilayers, Proc. Natl. Acad. Sci. USA 71 (1974) 1564–1568.
[33] F.T. Hong, D. Mauzerall, Tunable voltage clamp method: application to photoelectric effects in pigmented bilayer lipid membranes, J. Electrochem. Soc. 123 (1976) 1317–1324.
[34] H.-M. Ullrich, H. Kuhn, Photoelectric effects in bimolecular lipid-dye membranes, Biochim. Biophys. Acta 266 (1972) 584–596.
[35] H.-W. Trissl, Light-induced conformational changes in cattle rhodopsin as probed by measurements of the interfacial potential, Photochem. Photobiol. 29 (1979) 579–588.
[36] D. Oesterhelt, W. Stoeckenius, Rhodopsin-like protein from the purple membrane of *Halobacterium halobium*, Nat. New Biol. 233 (1971) 149–152.
[37] W. Stoeckenius, The purple membrane of salt-loving bacteria, Sci. Am. 234(6) (1976) 38–46.
[38] W. Stoeckenius, R.A. Bogomolni, Bacteriorhodopsin and related pigments of Halobacteria, Ann. Rev. Biochem. 51 (1982) 587–616.
[39] R. Henderson, J.M. Baldwin, T.A. Ceska, F. Zemlin, E. Beckmann, K.H. Downing, Model for the structure of bacteriorhodopsin based on high-resolution electron cryomicroscopy, J. Mol. Biol. 213 (1990) 899–929.
[40] H. Luecke, B. Schobert, H.-T. Richter, J.-P. Cartailler, J.K. Lanyi, Structural changes in bacteriorhodopsin during ion transport at 2 Angstrom resolution, Science 286 (1999) 255–260.
[41] J.K. Lanyi, Progress toward an explicit mechanistic model for the light-driven proton pump, bacteriorhodopsin, FEBS Lett. 464 (1999) 103–107.
[42] J.K. Lanyi, Bacteriorhodopsin, Biochim. Biophys. Acta 1460 (2000) 1–3.
[43] J.K. Lanyi, H. Luecke, Bacteriorhodopsin, Curr. Opin. Struct. Biol. 11(4) (2001) 415–419.
[44] J.K. Lanyi, A. Pohorille, Proton pumps: mechanism of action and applications, Trends Biotechnol. 19 (2001) 140–144.
[45] H. Luecke, J.K. Lanyi, Structural clues to the mechanism of ion pumping in bacteriorhodopsin, Adv. Protein Chem. 63 (2003) 111–130.
[46] J.K. Lanyi, Bacteriorhodopsin, Annu. Rev. Physiol. 66 (2004) 665–688.

[47] J.K. Lanyi, Proton transfers in the bacteriorhodopsin photocycle, Biochim. Biophys. Acta 1757 (2006) 1012–1018.
[48] T. Hirai, S. Subramaniam, J.K. Lanyi, Structural snapshots of conformational changes in a seven-helix membrane protein: lessons from bacteriorhodopsin, Cur. Opin. Struct. Biol. 19(4) (2009) 433–439.
[49] F.T. Hong, Retinal proteins in photovoltaic devices, in: R.R. Birge (Ed.), Molecular and Biomolecular Electronics, American Chemical Society, Washington, DC, 1994, pp. 527–559(Advances in Chemistry Ser. 240).
[50] F.T. Hong, Biomolecular electronics, in: R.F. Taylor, J.S. Schultz (Eds.), Handbook of Chemical and Biological Sensors, Institute of Physics Publishing Ltd., Bristol, Philadelphia, 1996, pp. 257–286.
[51] N. Vsevolodov, Biomolecular Electronics: An Introduction via Photosensitive Proteins, Birkhäuser, Boston, 1998.
[52] N. Hampp, Bacteriorhodopsin as a photochromic retinal protein for optical memories, Chem. Rev. 100 (2000) 1755–1776.
[53] N. Hampp, Bacteriorhodopsin: mutating a biomaterial into an optoelectronic material, Appl. Microbiol. Biotechnol. 53 (2000) 633–639.
[54] J.A. Stuart, R.R. Birge, P. Bhattacharia, B.J. Yordy, J. Girard, W. Tetley, et al. Bacteriorhodopsin: from biophotonic material to chemical sensor, in: G.K. Knopf, A.S. Bassi (Eds.), Smart Biosensor Technology, CRC Press, Boca Raton, 2007, pp. 355–384.
[55] F.T. Hong, Photoelectric biosensors: fundamentals and innovative designs, in: G.K. Knopf, A.S. Bassi (Eds.), Smart Biosensor Technology, CRC Press, Boca Raton, 2007, pp. 385–435.
[56] L. Lensu, M. Frydrych, J. Parkkinen, S. Parkkinen, T. Jaaskelainen, Color-sensitive biosensors for imaging applications, in: G.K. Knopf, A.S. Bassi (Eds.), Smart Biosensor Technology, CRC Press, Boca Raton, 2007, pp. 437–460.
[57] W.W. Wang, G.K. Knopf, A.S. Bassi, Protein-based photoreceptor array on flexible plastic substrates, in: G.K. Knopf, A.S. Bassi (Eds.), Smart Biosensor Technology, CRC Press, Boca Raton, 2007, pp. 461–502.
[58] K.T. Brown, M. Murakami, A new receptor potential of the monkey retina with no detectable latency, Nature 201 (1964) 626–628.
[59] R.A. Cone, W.L. Pak, The early receptor potential, in: W.R. Loewenstein (Ed.), Handbook of Sensory Physiology, Vol. 1, Principles of Receptor Physiology. Springer-Verlag, Berlin, 1971, pp. 345–365.
[60] F.T. Hong, M. Montal, Bacteriorhodopsin in model membranes: a new component of the displacement photocurrent in the microsecond time scale, Biophys. J. 25 (1979) 465–472.
[61] W.A. Hagins, R.E. McGaughy, Molecular and thermal origins of fast photoelectric effects in the squid retina, Science 157 (1967) 813–816.
[62] R.G. Matthews, R. Hubbard, P.K. Brown, G. Wald, Tautomeric forms of metarhodopsin, J. Gen. Physiol. 47 (1963) 215–240.
[63] M.A. Ostrovsky, Animal rhodopsin as a photoelectric generator, in: F.T. Hong (Ed.), Molecular Electronics: Biosensors and Biocomputers, Plenum Press, New York, London, 1989, pp. 187–201.
[64] D. Emeis, H. Kühn, J. Reichert, K.P. Hofmann, Complex formation between metarhodopsin II and GTP-binding protein in bovine photoreceptor membranes leads to a shift of the photoproduct equilibrium, FEBS Lett. 143 (1982) 29–34.
[65] F.T. Hong, Mechanisms of generation of the early receptor potential revisited, Bioelectrochem. Bioenerg. 5 (1978) 425–455.
[66] L.A. Drachev, A.D. Kaulen, L.V. Khitrina, V.P. Skulachev, Fast stages of photoelectric processes in biological membranes. I. Bacteriorhodopsin, Eur. J. Biochem. 117 (1981) 461–470.

[67] T.L. Okajima, F.T. Hong, Kinetic analysis of displacement photocurrents elicited in two types of bacteriorhodopsin model membranes, Biophys. J. 50 (1986) 901–912.
[68] F.T. Hong, Bacteriorhodopsin as an intelligent material, in: C.-Y. Wang, C.-T. Chen, C.-K. Cheng, Y.-Y. Huang, F.-H. Lin (Eds.), Biomedical Engineering in the 21st Century, National Taiwan University Center for Biomedical Engineering, Taipei, Taiwan, 1990, pp. 85–95.
[69] J. Deisenhofer, H. Michel, The photosynthetic reaction center from the purple bacterium *Rhodopseudomonas viridis*, Science 245 (1989) 1463–1473.
[70] J. Barber, Photosynthetic electron transport in relation to thylakoid membrane composition and organization, Plant Cell Environ. 6 (1983) 311–322.
[71] H.-W. Trissl, M. Montal, Electrical demonstration of rapid light-induced conformational changes in bacteriorhodopsin, Nature 266 (1977) 655–657.
[72] S. Michaile, F.T. Hong, Component analysis of the fast photoelectric signal from model bacteriorhodopsin membranes: part I. Effect of multi-layer stacking and prolonged drying, Bioelectrochem. Bioenerg. 33 (1994) 135–142.
[73] L.A. Drachev, A.D. Kaulen, V.P. Skulachev, Time resolution of the intermediate steps in the bacteriorhodopsin-linked electrogenesis, FEBS Lett. 87 (1978) 161–167.
[74] T.L. Okajima, S. Michaile, L.E. McCoy, F.T. Hong, Component analysis of the fast photoelectric signal from model bacteriorhodopsin membranes: part II. Effect of fluorescamine treatment, Bioelectrochem. Bioenerg. 33 (1994) 143–149.
[75] F.H. Hong, M. Chang, B. Ni, R.B. Needleman, F.T. Hong, Component analysis of the fast photoelectric signal from model bacteriorhodopsin membranes: part III. Effect of the point mutation aspartate 212 $\rightarrow$ asparagine 212, Bioelectrochem. Bioenerg. 33 (1994) 151–158.
[76] S. Michaile, A. Duschl, J.K. Lanyi, F.T. Hong, Chloride ion modulation of the fast photoelectric signal in halorhodopsin thin films, in: B. Oranal, P.C. Pedersen (Eds.), Proceedings of the 12th Annual International Conference of IEEE Engineering in Medicine and Biology Society, Philadelphia, PA, November 1–4, 1990, Institute of Electrical and Electronics Engineers, Washington, DC, 1990, pp. 1721–1723.
[77] T. Miyasaka, K. Koyama, I. Itoh, Quantum conversion and image detection by a bacteriorhodopsin-based artificial photoreceptor, Science 255 (1992) 342–344.
[78] J.P. Wang, J.R. Li, P.D. Tao, X.C. Li, L. Jiang, Photoswitch based on bacteriorhodopsin Langmuir-Blodgett films, Adv. Mater. Optics Electron. 4 (1994) 219–224.
[79] K. Koyama, N. Yamaguchi, T. Miyasaka, Antibody-mediated bacteriorhodopsin orientation for molecular device architectures, Science 265 (1994) 762–765.
[80] F.T. Hong, Molecular sensors based on the photovoltaic effect of bacteriorhodopsin: origin of differential responsivity, Mater. Sci. Eng. C5 (1997) 61–79.
[81] L.A. Drachev, A.A. Jasaitis, A.D. Kaulen, A.A. Kondrashin, E.A. Liberman, I.B. Nemecek, et al. Direct measurement of electric current generation by cytochrome oxidase, H-ATPase and bacteriorhodopsin, Nature 249 (1974) 321–324.
[82] D. Mauzerall, A. Finkelstein, Light induced changes in the conductivity of thin lipid membranes in the presence of iodine and iodide ion, Nature 224 (1969) 690–692.
[83] J. Kutnik, Z. Łojewska, Photoconductivity of bilayer lipid membranes with chlorophyll and carotenoids, Stud. Biophys. 82 (1981) 127–135.
[84] C.M. Drain, B. Christensen, D. Mauzerall, Photogating of ionic currents across a lipid bilayer, Proc. Natl. Acad. Sci. USA 86 (1989) 6959–6962.
[85] F.T. Hong, D. Mauzerall, The separation of voltage-dependent photoemfs and conductances in Rudin-Mueller membranes containing magnesium porphyrins, Biochim. Biophys. Acta 275 (1972) 479–484.
[86] B.E. Fuller, T.L. Okajima, F.T. Hong, Analysis of the d.c. photoelectric signal from model bacteriorhodopsin membranes: d.c. photoconductivity determination by means of the null current method and the effect of proton ionophores, Bioelectrochem. Bioenerg. 37 (1995) 109–124.

[87] T.R. Hesketh, Photoconductivity in black lipid and thin lipid membranes, Nature 224 (1969) 1026–1028.

[88] Z. Dancsházy, B. Karvaly, Incorporation of bacteriorhodopsin into a bilayer lipid membranes: a photoelectric-spectroscopic study, FEBS Lett. 72 (1976) 136–138.

[89] E. Bamberg, H.-J. Apell, N.A. Dencher, W. Sperling, H. Stieve, P. Läuger, Photocurrents generated by bacteriorhodopsin on planar bilayer membranes, Biophys. Struct. Mech. 5 (1979) 277–292.

[90] T.R. Herrmann, G.W. Rayfield, The electrical response to light of bacteriorhodopsin in planar membranes, Bioshys. J. 21 (1978) 111–125.

[91] L.A. Drachev, V.N. Frolov, A.D. Kaulen, E.A. Liberman, S.A. Ostroumov, V. G. Plakunova, et al. Reconstitution of biological molecular generators of electric current: bacteriorhodopsin, J. Biol. Chem. 251 (1976) 7059–7065.

[92] E. Bamberg, N.A. Dencher, A. Fahr, M.P. Heyn, Transmembraneous incorporation of photoelectrically active bacteriorhodopsin in planar lipid bilayers, Proc. Natl. Acad. Sci. USA 78 (1981) 7502–7506.

[93] V.M. Mirsky, V.S. Sokolov, T.V. Dyukova, E.I. Melnik, A study of bacteriorhodopsin-containing proteoliposome incorporation into bimolecular lipid-membranes, Bioelectrochem. Bioenerg. 11 (1983) 327–346.

[94] P. Seta, P. Ormos, B. D'Ependoux, C. Gavach, Photocurrent response of bacteriorhodopsin adsorbed on bimolecular lipid membranes, Biochim. Biophys. Acta 591 (1980) 37–52.

[95] C. Gavach, P. Seta, E. Bienvenue, Photogenerated movement of protons through bacteriorhodopsin in relation to the analysis of photoelectrical responses, Faraday Discuss. Chem. Soc. 74 (1982) 365–375.

[96] G. Szabó, E. Bamberg, Current voltage relationship for the light driven proton pump bacteriorhodopsin, Biophys. J. 47 (1985) 332a.

[97] G. Nagel, B. Möckel, G. Büldt, E. Bamberg, Functional expression of bacteriorhodopsin in oocytes allows direct measurement of voltage dependence of light-induced H^+ pumping, FEBS Lett. 377 (1995) 263–266.

[98] G. Nagel, B. Kelety, B. Möckel, G. Büldt, E. Bamberg, Voltage dependence of proton pumping by bacteriorhodopsin is regulated by voltage-sensitive ratio of M_1 to M_2, Biophys. J. 74 (1998) 403–412.

[99] W.J. Vredenberg, A.A. Bulychev, Changes in the electrical potential across the thylakoid membranes of illuminated intact chloroplasts in the presence of membrane-modifying agents, Plant Sci. Lett. 7 (1976) 101–107.

[100] N.K. Packham, P. Mueller, P.L. Dutton, Photoelectric currents across planar bilayer membranes containing bacterial reaction centers: the response under conditions of multiple reaction-center turnovers, Biochim. Biophys. Acta 933 (1988) 70–84.

[101] E.L. Barsky, Z. Dancsházy, L.A. Drachev, M.D. Il'ina, A.A. Jasaitis, A.A. Kondrashin, et al. Reconstitution of biological molecular generators of electric current. Bacteriochlorophyll and plant chlorophyll complexes, J. Biol. Chem. 251 (1976) 7066–7071.

[102] R.J.P. Williams, Possible functions of chains of catalysts, J. Theor. Biol. 1 (1961) 1–17.

[103] P. Mitchell, Chemiosmotic coupling in oxidative and photosynthetic phosphorylation, Biol. Rev. 41 (1966) 445–502.

[104] A.E. Blaurock, W. Stoeckenius, Structure of the purple membrane, Nat. New Biol. 29 (1971) 152–155.

CHAPTER FIVE

A Multiparametric Fluorescence Approach for Biomembrane Studies

Ahmed A. Heikal[1,2,*]

Contents

Abstract

Biological membranes are heterogeneous assemblies of a variety of lipids and proteins as well as cholesterol. The dynamic nature of these biomembranes spans a wide range of spatiotemporal scales that are essential to their function in cell signaling and biomolecular trafficking. In contrast, biomimetic membranes are simple, stable, and versatile systems to study the physicochemical principles such as lipid mixing, phase separation, domain formation and intermolecular interactions, which underlie lipid bilayers assemblies under controlled conditions. In this chapter, multiparametric fluorescence micro-spectroscopy approach will be described for quantitative and noninvasive investigation of molecular processes that trigger lateral and temporal heterogeneities in biomembranes.

* Corresponding author. Tel.: +1-218-726-7036; Fax: +1-218-726-7394.
E-mail address: aaheikal@d.umn.edu

[1] The Department of Chemistry and Biochemistry, Swenson College of Science and Engineering, Duluth, Minnesota, USA

[2] The Department of Pharmacy Practice and Pharmaceutical Sciences, College of Pharmacy, The University of Minnesota Duluth, Duluth, Minnesota, USA

Advances in Planar Lipid Bilayers and Liposomes, Volume 13
ISSN 1554-4516, DOI: 10.1016/B978-0-12-387721-5.00005-5

1. Introduction

Biological membranes play important roles in cell signaling and biomolecular trafficking [1,2]. They consist of heterogeneous and dynamic assemblies of lipids, proteins and cholesterol that are organized into asymmetric bilayers [3]. The polar hydrophilic headgroups of phospholipids in membranes are exposed to the aqueous environment while the hydrocarbon tails constitute the hydrophobic core. The lateral heterogeneity of cholesterol content in these biomembranes causes the lipids and membrane proteins to segregate into lipid domains with distinct biophysical properties. Based on their detergent-resistance in biochemical assays, liquid-ordered domains (or "lipid rafts" [4]) were first proposed to exist in the plasma membrane at low temperatures [5]. These domains are postulated to be enriched in phospholipids and glycosphingolipids as well as cholesterol and GPI-anchored proteins [6–8]. However, lipid rafts remain elusive [9] in living cells under physiological conditions, perhaps due to their small sizes and/or transient nature. Two-dimensional Ising model analysis [10] and single-particle tracking in a laser trap [6] of lipid rafts in intact cells suggest sizes of 20 nm and 26 ± 13 nm, respectively, which is well below the diffraction limit of conventional microscopy [11]. Yet, these specialized liquid-ordered domains have been implicated in a number of essential biological processes such as ligand-activated receptor signaling and viral infection [12,13].

In contrast, planar and vesicular model membranes provide a simple and versatile system to investigate intermolecular interactions and large domain formation under controlled lipid compositions [1,14–21]. The composition of these model membranes can vary from single lipid components to mixtures of a few lipids (either synthetic or natural extracts) and cholesterol [4]. In planar model membranes, for example, phospholipids are usually supported on the air–water interface that forms a single monolayer or free standing bilayers, where the bottom layer may be anchored to the substrate via a polymer cushion [3]. These planar models have been used for a wide range of membrane studies such as surface tension, membrane conductivity, channel formation, and interleaflet coupling [3,22–25]. Alternatively, vesicular model membranes (liposomes) are prepared with multilamellar (multiple bilayers) or unilamellar (single bilayer) vesicles, which exhibit a significant variance in sizes such as small (SUVs, starting at ~20 nm), large (LUVs, 100–200 nm), and giant (GUVs, 10–100 μm) unilamellar vesicles. The large size of LUVs and GUVs leads to a reduced curvature stress and therefore more stability during optical microscopy imaging. GUVs [26] provide a free standing bilayer (i.e., no substrate effect) that have been widely used in numerous studies related to lipid dynamics and raft formation, vesicle shape changes, membrane tube formation and membrane fusion. As a result, these models are ideal systems for phase segregation, lipid–protein and protein–protein interactions, vesicular shape changes, and membrane permeability studies [27–31]. Lipids in model membranes can laterally segregate into coexisting liquid-ordered (L_o) and liquid-disordered (L_d) domains based on the lipid type (saturated vs. unsaturated), cholesterol content, and temperature [32]. Single phase GUVs that are made from saturated lipids (e.g., sphingomyelin [SM]) exhibit a highly ordered gel phase (L_β) [33,34]. Liquid-ordered lipid domains have attracted a particular attention because they are believed to be biologically relevant with a potential role in various cellular mechanisms including signal transduction, cellular transport, and membrane fusion [35–37]. Based on lipid mixing, the ability to predict the formation of specific lipid domains is of particular interest [38]. Supported lipid bilayers allow for the formation of asymmetric bilayers that are difficult to achieve using GUVs. The geometrical features of these planar bilayers allow for a relatively easy analysis of the size and area of lipid domains [39].

Biomimetic membranes also provide a platform for investigating the effects of harmful peptides or toxins to the integrity of biological membranes. The underlying mechanisms of such membrane destabilization include membrane solubilization (i.e., peptides acting as detergents) and/or pore formation across the bilayer via leakage experiments [39]. Incorporating proteins into model bilayers is particularly important in testing the hypothesis that membrane proteins emulsify lipid domains in intact cell membranes leading to smaller and more dynamic domains [40,41]. Hinderliter *et al.* demonstrated that changing the chemical structure of lipids will affect the lipid–lipid interactions while altering protein-induced lipid-domain formation in a predictable manner [41]. One of the challenges in incorporating membrane proteins into model bilayers, however, is that most amphiphilic membrane proteins denature in the organic solvents used in vesicles preparation. Yet, small proteoliposomes containing bacteriorhodopsin have been successfully fused with GUVs [42–44]. Doeven *et al.* used sucrose to prevent the inactivation of distinct membrane proteins during the dehydration step of the GUVs formation process [45]. Another new method was reported for reconstituting membrane proteins (sarcoplas mic reticulum Ca^{2+}-ATPase and the proton pump bacteriorhodopsin) into GUVs [46].

There are some concerns, however, about the biological relevance of model membranes [1,4,47]. At the center of this ongoing discussion is the stable large-sized domains observed in model membrane that remain elusive in the plasma membrane of live cells under physiological conditions. These thermodynamically stable lipid domains in model membranes are fundamentally different from the dynamic nature of biological membranes. In addition, model membranes are mostly made from one or two lipids as a function of cholesterol content as compared with the large variety of lipid types and membrane proteins present in biological membranes. The large variation of lipids in biological membranes minimizes line tension, which can be a contributing factor in creating large domains in biomimetic membranes. To address some of these concerns, giant plasma membrane vesicles (GPMVs), also known as membrane blebs, were introduced [48] as an alternative model. GPMVs (12 ± 8 μm in diameter) are isolated from the plasma membrane of mammalian cells with lipid variation and membrane proteins intact, which provide a more realistic biomimetic model [49]. However, the cytoskeleton underlying the plasma membrane in intact cells is absent from the GPMVs model. Recent studies demonstrated the presence of lipid domains in these GPMVs at low temperature [11,49,50], which resemble those domains observed in GUVs. In addition, transmembrane and outer leaflet anchored proteins are generally found to partition in liquid-ordered like domains. In contrast, the inner leaflet anchored proteins are shown to partition in the liquid-disordered phase [49]. This liquid–liquid phase separation in GPMVs at low temperature, however, is absent from the plasma membrane of mammalian cells under either similar or physiological conditions. One contributing factor may be the absence of the actin cytoskeleton that may regulate the lipid heterogeneity in the plasma membrane in living cells [51,52]. The size and lifetime of lipid domains in GPMVs fluctuate as the temperature approaches the transition temperature of phase separation [10].

Biophysical methods play a central role in lipid domain studies and other biophysical aspects of biomembranes. This chapter describes a multimodal fluorescence micro-spectroscopy approach (Fig. 1) to study molecular organization, interaction, diffusion and order in a lipid bilayer. In this multiparametric approach, the thermodynamic variables of membrane assembly include lipid type, cholesterol content, protein presence, temperature, and mechanical forces, which can be controlled. The biophysical questions of interest will then dictate what model membrane to use as well as the nature of the thermodynamic quantities to be manipulated. The fluorescence (confocal or two-photon) microscope is used to visualize the formation and dynamics of lipid domains under specific thermodynamics conditions. Variation of the membrane local environment (i.e., lateral heterogeneity), lipid domains, and the interactions among the constituents of bilayers can be quantified using fluorescence lifetime imaging microscopy (FLIM) due to its superior sensitivity to structural and environmental changes. The order of lipid analogs in the bilayer as well as the membrane fluidity is quantified in space and time using fluorescence polarization imaging and time-resolved anisotropy measurements, respectively. Finally, fluorescence correlation spectroscopy (FCS) is the tool of choice for measuring the diffusion kinetics of lipid analogs and proteins in the membrane. This chapter is not intended to be a thorough review of membrane studies,

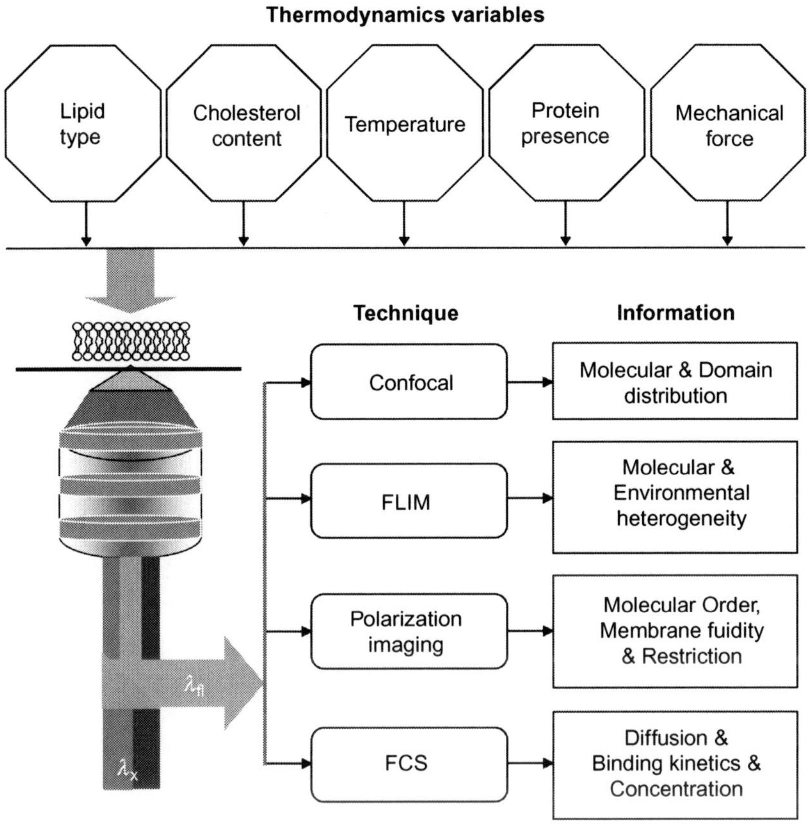

Figure 1 Multiparametric approach for studying different biophysical aspects of lipid domains in model membranes. From the model membrane perspective, a range of thermodynamics parameters can be controlled for creating distinct lipid domains according to a phase diagram. These thermodynamic parameters include lipid type, cholesterol content, temperature, proteins, and mechanical forces. The type of biophysical studies will dictate the experimental design using the proposed multiparametric fluorescence approach. For example, the excitation and detection wavelength will depend on the fluorescent lipid analogs being used for lipid domain imaging using confocal (one-photon, blue) or two-photon (red) microscopy. Complementary fluorescence lifetime imaging (FLIM) exploits the sensitivity of fluorescence lifetime of lipid analogs to different domains and the lateral heterogeneity of the membrane. In addition, fluorescence lifetime and intensity images, recorded simultaneously on a calibrated microscope, can be used to estimate the concentration of fluorescent lipids. Polarization-analyzed imaging of lipid domains is also used to quantify the order parameter and fluidity of lipid domains. FCS is another essential tool for fluctuation analysis and the quantification of lateral diffusion kinetics of a lipid analog in a lipid domain. The proposed multiparametric approach provides rich biophysical information concerning the formation and dynamics of lipid domains under controlled thermodynamics. (For interpretation of the references to color in this figure legend, the reader is referred to the Web version of this chapter.)

but rather to serve as a technical resource for researchers who are interested in different aspects of membrane biophysics. In so doing, recent developments in model membrane studies will be selectively highlighted within the context of each technique.

2. Visualization of Lipid Domains in Model Membranes

Confocal and two-photon fluorescence microscopy has been used to investigate the segregation of lipids into phase domains as a function of lipid type and cholesterol content in model membranes. The general approach for visualizing lipid domains in model membranes involves the use of phase specific fluorescent markers [39]. Multichannel confocal images are then recorded using the appropriate excitation laser and fluorescence emission wavelengths for each label. While the excitation laser in these experiments is polarized, the emitted fluorescence is depolarized. Figure 2 shows different DIC (differential interference contrast) and confocal images of lipid domains in GUVs, labeled with DiI-C_{12} (liquid-disordered) and Alexa-cholera-toxin B (Alexa-CTXB, liquid-ordered). Two-dimensional DIC (Fig. 2A) and confocal (Fig. 2B) cross-section of a single GUV indicate the localization of DiI-C_{12} in the lipid bilayer of fluid GUV made of dioleylphosphatidylcholine (DOPC). The corresponding three-dimensional confocal image (Fig. 2C) shows a homogenous distribution of DiI-C_{12} on the lipid bilayer of a vesicle. In ternary GUVs, liquid-ordered (Fig. 2D) and liquid-disordered lipid domains coexist under specific lipid mixing, shown here as a cross-section of a single vesicle labeled with Alexa-CTXB and DiI-C_{12}, respectively. Two-channel three-dimensional confocal imaging of liquid–liquid phase coexistence in ternary GUVs is also shown (Fig. 2G and H) using 1:1 of DOPC and SM lipids with 20% mol of cholesterol. Different shapes of liquid–liquid domains (Fig. 2I) can be created with controlled lipid and cholesterol mixing [32–34]. Panels (H)–(I) in Fig. 2 demonstrate the segregation of DiI-C_{12} and Alexa-CTXB in the coexisting liquid–liquid lipid domains of ternary GUVs.

Ariola *et al.* have recently investigated the partitioning and dynamics of a new fluorescent cholesterol analog (namely, Bodipy-Cholesterol) in ternary-phase GUVs [33]. In those studies, the GUVs were made from different molar ratios of DOPC, SM, and cholesterol. The specific lipid compositions were determined from the published phase diagram for DOPC/SM/cholesterol in GUVs at 23 °C [53]. The naturally fluorescent dehydroegosterol (DHE) is another cholesterol analog that induces liquid-ordered domains in GUVs [54]. In contrast with the nonfluorescent cholesterol, the DHE fluorescence originates from two additional double bonds in

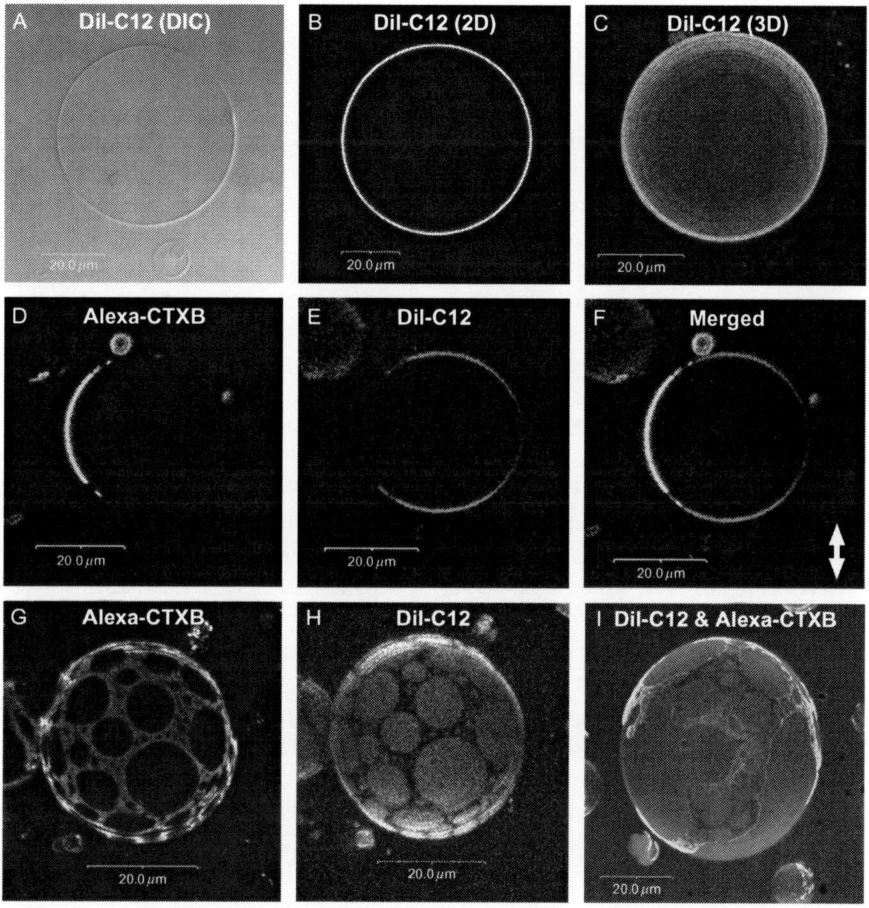

Figure 2 Two- and three-dimensional imaging of lipid phases in GUVs using confocal microscopy. Optical cross-sectioning, using DIC (A) and confocal (B: 2D, C: 3D) microscopy reveal that DiI-C_{12} is homogeneously distributed and localized in the lipid bilayer of a single DOPC phase, liquid-disordered GUV. In ternary GUVs, liquid-ordered (D) and liquid-disordered (E) domains are labeled with Alexa-cholera-toxin B (Alexa-CTXB) and DiI-C_{12}, respectively, as shown using 2D optical sectioning. The merged liquid–liquid lipid phases (F) indicates their laterally heterogeneous coexistence in the same GUVs. In the region of liquid–liquid coexistence of the phase diagram, 3D imaging using DiI-C_{12} and Alexa-CTXB reveals interesting shapes (G–I) of ternary GUVs. These images were recorded using 543-nm excitation with 1.2 *NA*, water immersion objective.

the steroid ring system. However, this cholesterol analog absorbs in the UV, which requires special optics. Baumgart *et al.* also reported on the liquid–liquid domain coexistence in GUVs model with a direct correlation between the lipid composition, local membrane curvature, and line tension [32]. In addition, a long-range domain ordering was observed in the form of locally parallel stripes and hexagonal arrays of circular lipid domains, curvature-dependent domain sorting, and membrane fission into separate vesicles at domain boundaries [32].

Two-photon excitation microscopy [55–57] has some advantages over confocal microscopy in studying lipid-domain formation [35,58,59]. For example, the spatial resolution in two-photon microscopy is inherently high (due to the nonlinear excitation) without the need for de-scanning and a pinhole for rejecting out-of-focus photons, which are keys in confocal microscopy. It is also possible to excite multiple phase specific fluorophores for monitoring lipid-domain formation. Importantly, the excitation wavelength in two-photon microscopy (730–1000 nm) is easily distinguishable from the fluorescence emission wavelength ($\sim$400–650 nm), which minimizes the detection of laser scattering. A typical two-photon fluorescence image of DiI-C_{12} in single membrane vesicle, isolated from the plasma membrane of an epithelial HTB126 cell, is shown in Fig. 3A. The angular distribution of fluorescence intensity can be understood in terms of the relative orientation of the DiI-C_{12} dipole with respect to the laser polarization (arrow), especially under polarization-analyzed emission. The morphology of lipid domains in GUVs has also been characterized using two-photon microscopy imaging of 6-dodecanoyl-2-(dimethylamino) naphthalene (Laurdan) at different temperatures [35] as well as a newly developed cholesterol analog (namely, Bodipy-cholesterol) [33]. Guided by the phase diagram, lipid domains of different sizes and shapes were also observed. In another study, di-4-ANEPPDHQ was found to partition into both liquid-ordered and liquid-disordered phase domains in model membranes using both linear and nonlinear imaging [60]. The dye is water soluble and yet it exhibits a high affinity to lipid membranes. Importantly, the fluorescence of di-4-ANEPPDHQ is blue-shifted by 60 nm in liquid-ordered phase as compared with liquid-disordered phases. Interestingly, di-4-ANEPPDHQ has generated a greater second harmonic signal in the liquid-disordered phase.

3. Fluorescence Lifetime Imaging is Sensitive to Lipid Phase and Intermolecular Interactions

Lipid–lipid and lipid–protein interactions play a critical role in membrane assembly and lipid-phase segregation into domains. The restriction of these membrane constituents in two-dimensional bilayers enhances such intermolecular interaction probability, which leads to lateral heterogeneity in local concentrations, bilayer thickness, and lateral pressure in biomembranes. Such heterogeneity is likely to regulate the function of membrane

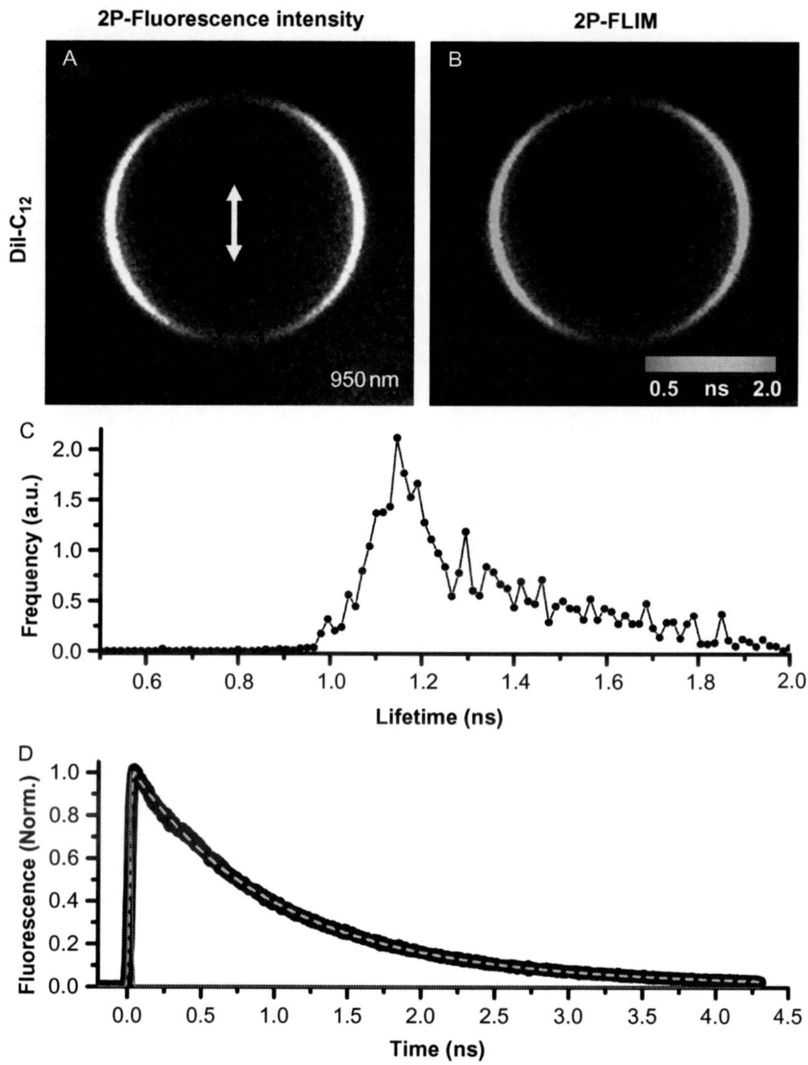

Figure 3 Two-photon fluorescence lifetime image of DiI-C_{12}-labeled GPMVs isolated from HTB126 epithelial cells. The two-photon fluorescence intensity of DiI-C_{12} (A) was excited using 950 nm in a single vesicle. The corresponding two-photon fluorescence lifetime image is also shown for the same vesicle (B). The fluorescence emission in these measurements was detected at the magic angle to remove any rotational effect on the excited state dynamics of DiI-C_{12}. As a result, the polarization-analyzed excitation and detection of DiI-C_{12} emission leads to the observed angular-dependence of the fluorescence with respect to the laser polarization (the arrow), which is directly related to the relative orientation of the DiI-C_{12} dipole in the membrane. The color code (B) and the pixel-lifetime histogram (C) reflects the average lifetime distribution of the DiI-C_{12} in each pixel of this image. Complementary

proteins in cell signaling [61]. Due to the sensitivity of fluorescence lifetime to molecular structure, intermolecular interactions, and local environment, FLIM is a powerful tool for elucidating the underlying mechanisms for the lateral heterogeneity of biomembranes. What makes FLIM so appealing is that the same fluorescence label can be used to probe different lipid domains. To do so, however, an understanding of the spectroscopy and excited-state dynamics of a given label is required.

Unlike confocal microscopy, two-photon FLIM of biomolecules is mostly carried out using laser pulses for excited-state interrogation. The detection of excited state fluorescence photons, emitted by a fluorophore in a given environment, is a statistical process that is generally described as an exponential decay. The complexity of such fluorescence decay depends on a number of molecular and environmental parameters such as the chemical structure of the fluorophore, temperature, *p*H, the local environment, and competing nonradiative processes of the excited state. Hence, FLIM method is uniquely sensitive to the chemical structure of lipid analogs and proteins, the local environment, and the manner in which they are intercalated with the membrane. There are two modes of operation for FLIM in either time or frequency domains, which are described in details elsewhere [62]. The discussion here will be limited to time-domain FLIM [63], which is based on time-correlated single-photon counting (TCSPC) [63,64].

In TCSPC, the fluorophore is exciting with a laser pulse and, simultaneously, the electronics are triggered (or synchronized). The arrival time of the first detected fluorescence photon is recorded for each excitation-detection cycle. The probability histogram of detecting a fluorescence photon per cycle is recoded as a function of time, which constitutes the fluorescence decay that describes the excited-state dynamics of a given fluorophore. This excitation-detection cycle is repeated to enhance the signal-to-noise ratio, which is critical for multiple exponential decay analysis. However, these fluorescence decays are convoluted with the systems response function and therefore a deconvolution algorithm is required to extract the exact excited state dynamics, especially when both take place on comparable time scales. Based on the chemical structure and surrounding environment of a given fluorophore, the fluorescence emission is generally depolarized due to rotational diffusion during its excited state lifetime. As a result, the fluorescence emission should be detected at the magic angle

one-photon, single-point measurements of the DiI-C_{12} fluorescence in similar GPMVs (D) provide reliable data with good signal-to-noise level and high temporal resolution as a point of reference for FLIM. The fluorescence decay is a biexponential with $\tau_1 = 1.0 \pm 0.1$ ns, $\alpha_1 = 0.79 \pm 0.09$, $\tau_2 = 1.8 \pm 0.3$ ps, and $\alpha_2 = 0.2 \pm 0.1$) with an estimated average lifetime of 1.1 ± 0.1 ns. The fitting curve (dotted, gray) and the system response function are also shown (D). (For interpretation of the references to color in this figure legend, the reader is referred to the Web version of this chapter.)

(54.7° with respect to the excitation laser polarization) to eliminate any rotational effect on the excited state dynamics [63–65]. Based on the chromophore structure and surrounding lipid phase, the magic-angle fluorescence intensity decay ($I_{54.7}$) of a chromophore can be generally described as [34,66–68]

$$I_{54.7}(x,y,t) = \sum_{i=1}^{3} \alpha_i(x,y)\exp[-t/\tau_i(x,y)]. \quad (1)$$

The time constants (τ_i) and amplitudes (α_i) are used to calculate the average fluorescence lifetime, $\sum \alpha_i\tau_i/\alpha_i$. These measurements can be done in either scanning (FLIM) or single-point mode, where the 1P excitation laser is strategically focused on a selected lipid domain [33]. The complementary single-point measurements provide both high signal-to-noise ratio as well as temporal resolution as compared to FLIM. However, FLIM image provides superior spatial resolution that can be advantageous for phase separation and lipid-domain formation studies.

A typical cross-sectional FLIM image of a DiI-C_{12}-labeled GPMV, isolated from HTB126 epithelial cells, is shown in Fig. 3B under 950-nm illumination. The fluorescence per pixel decays as a biexponential with $\tau_1 = 0.9 \pm 0.2$ ns (0.8 ± 0.1) and $\tau_2 = 2.1 \pm 0.3$ ps (0.2 ± 0.1), and an estimated average lifetime of 1.13 ± 0.09 ns ($n = 12$ pixels). The integrated two-photon fluorescence intensity per pixel is also shown (Fig. 3A) for the same vesicle. The apparent angular dependence of the fluorescence intensity in these images is attributed to the polarization-analysis (and therefore the order) of the DiI-C_{12} dipole moment distribution (with respect to the laser polarization) within the lipid bilayer. The lifetime–pixel histogram (Fig. 3C) is broad with a peak around 1.13 ns and a full-width-half-maximum of 0.24 ns. With a negligible background, such broad distribution is likely due to a heterogeneous environment of the lipid bilayer. To enhance the signal-to-noise and temporal resolution in FLIM, a complementary single-point fluorescence decay of DiI-C_{12} in GPMV was also measured (see Fig. 3D), where the one-photon laser pulses are strategically focused on the membrane during data acquisition. At room temperature, the fluorescence decays as a biexponential ($\tau_1 = 1.0 \pm 0.1$ ns, $\alpha_1 = 0.79 \pm 0.09$, $\tau_2 = 1.8 \pm 0.3$ ps, and $\alpha_2 = 0.2 \pm 0.1$) with an estimated average lifetime of 1.1 ± 0.1 ns, which agrees with the mean value in FLIM image. Loura *et al.* also reported a biexponential fluorescence decays for DiI-C_{12} in the lipid bilayer of large unilamellar vesicles [69]. The measured average fluorescence lifetime of DiI-C_{12} in DLPC and DSPC LUVs were 0.74 and 1.06 ns, respectively [69]. The two fluorescence lifetime components of DiI-C_{12} also persist in membrane blebs isolated from HTB126 using either GPMV buffer or H_2O_2 treatment as well as in the lipid bilayer of fluid GUVs [34]. In single-phase, DOPC vesicles, the DiI-C_{12} fluorescence also decays as a biexponential with 0.75 ± 0.04 ns average lifetime [34]. These studies show that the fluorescence lifetime of DiI-C_{12} is sensitive to lipid type. Comparative studies on DiI-C_{12} in methanol also indicate a biexponential fluorescence decay with $\tau_1 = 67$ ps, $\alpha_1 = 0.11$, $\tau_2 = 254$ ps, $\alpha_2 = 0.89$, and an average lifetime 0.25 ± 0.03 ns [34]. This would suggest that the biexponential decay of DiI-C_{12} is attributed to the excited-state processes, which are sensitive to the surrounding environment. However, the biexponential fluorescence decays of DiI-C_{12} can also be attributed to structural and environmental heterogeneity of the local membrane as shown by time-resolved anisotropy measurements (see below). Using the same experimental approach, Ariola *et al.* demonstrated that the fluorescence lifetime of Bodipy-PC (i.e., tail-labeled lipid analog) is a sensitive probe for lipid order, particularly in the hydrophobic region of the bilayer of single-phase DOPC (L_d) and DPPC (L_β) vesicles [34].

Fluorescence lifetime measurements can also be used to calculate the efficiency of Förster resonance energy transfer (FRET) [70], which is also a powerful method to evaluate the lateral organization of lipids and their segregation into nanoscopic domains in the plasma membrane of live cells [67] or model membranes [33,69]. The spatial sensitivity of FRET to the donor–acceptor distance (≤ 10 nm) in the bilayer allow for defeating the diffraction limit ($\sim\lambda/2$) on the spatial resolution inherent in conventional optical microscopy. Importantly, such sensitivity to donor–acceptor distance is ideal for studying the elusive lipid domains "or rafts" in living cells, which are believed to be a few nanometers in size. In time-resolved FRET, the energy transfer efficiency ($E = 1 - \tau_{DA}/\tau_D$) can also be calculated using the measured fluorescence lifetime for the donor with (τ_{DA}) and without (τ_D) the presence of an acceptor [66]. FRET between two fluorophores can provide insights into donor and acceptor partitioning in lipid domains of a bilayer. For example, Ariola *et al.* have demonstrated recently that 2P-FLIM images of a new Bodipy-cholesterol analog are sensitive to lipid phase domains in ternary GUVs, especially in the presence of DiI-C_{12} as a label for the liquid-disordered phase [33]. Based on these results, it was concluded that Bodipy-cholesterol and DiI-C_{12} are a likely FRET pair for studying lateral heterogeneity in biomembranes. The estimated FRET efficiency in L_d domains is 0.19 ± 0.07 with an average donor–acceptor distance (R) of 2.6 ± 0.2 nm [33]. The fluorescence decay of the donor, in the presence of an acceptor, in lifetime-based FRET measurements, can also be used to estimate the partitioning coefficient of the donor in a given lipid phase in a lipid bilayer [71].

Like any other imaging techniques, FLIM has some drawbacks. For example, the signal-to-noise ratio per pixel is usually low due to the large number of pixels per image, which requires a long data acquisition time. To reduce the acquisition time in FLIM measurements, the image size can be reduced. Another approach to reduce the acquisition time is to reduce the number of time bins per pixel at the expense of the temporal resolution, which is usually low in FLIM measurements. The maximum achievable photon count rate is also limited to $\sim 10\%$ of the repetition rate of the laser pulses to avoid pile-up effect. Complementary single-point measurements complement the FLIM method with improved signal-to-noise ratio as well as temporal resolution [33,34,67,68]. The heterogeneity of biomolecular concentration in cells could also lead to regions with low S/N in FLIM images, which could complicate the analysis. Phasor approach simplifies FLIM image analysis to identify regions of similar lifetime [72–74] and therefore chemical composition as well as local environment. Finally, care must be taken when interpreting the multiexponential fluorescence decays, which can be attributed to heterogeneity in local environment, variation in the chemical compositions, or excited state processes of the same fluorophore. In addition to other control experiments, complementary time-resolved anisotropy measurements would enable us to examine whether the multiexponential fluorescence decays in FLIM can be attributed to structural (or environmental) heterogeneity of lipid domains [33,34,75].

4. Order and Fluidity of Lipid Bilayers as Measured Using Fluorescence Polarization Imaging

Lipid dynamics, lipid packing, and order fluctuations are critical for bilayer assembly, membrane fluidity, lipid-phase transition, and the coexistence of lipid domains. The dynamic nature of lipids regulates lipid–lipid, lipid–cholesterol, and lipid–protein interactions, which may lead to the transient formation of specialized domains. The corresponding lipid packing (fatty-acyl chain conformation) and order fluctuations are likely to play an important role in many membrane-mediated events such as passive Na^+ transport, membrane fluidity, and bilayer compressibility. Lipid dynamics take place over a wide range of time scales (picoseconds to microseconds) and, as a result, time-resolved fluorescence anisotropy [76] is a powerful tool for investigating lipid conformation dynamics and order fluctuation on these time scales. Complementary steady-state anisotropy imaging of a fluorescent lipid yields valuable information concerning its orientation, order, and fluidity in the lipid bilayer with respect to the excitation-laser polarization. The temperature dependence of steady-state fluorescence emission anisotropy of a lipid analog would also reveal the phase transition profiles of the lipid bilayer. These biophysical parameters of lipid bilayers are also sensitive to phase separation and, therefore, essential in modeling lipid compartmentalization [77].

Anisotropy imaging, $r(x, y, t)$, of a lipid- or cholesterol-analog, at a given pixel in a bilayer, can be calculated using the simultaneously measured parallel, $I_{||}(x, y, t)$, and perpendicular, $I_{\perp}(x, y, t)$, fluorescence polarization images such that [34,67,68]

$$r(x, y, t) = \frac{[I_{||}(x, y, t) - G \cdot I_{\perp}(x, y, t)]}{[I_{||}(x, y, t) + 2G \cdot I_{\perp}(x, y, t)]}. \tag{2}$$

The polarization-analyzed intensities must be corrected for the background signal. The G-factor accounts for possible polarization-biased detection efficiency and is estimated using the tail-matching approach [66]. The steady-state anisotropy can be calculated as the time (t) approaches infinity. The corresponding time-resolved anisotropy can generally be described as a multiexponential decay function. For example, time-resolved anisotropy of DiI-C_{12} in a lipid bilayer can be generally fitted to

$$r(x, y, t) = \sum_{i=1}^{2} \beta_i(x, y)\exp[-t/\phi_i(x, y)], \tag{3}$$

where the preexponential factor (β_i) represents the relative contribution of the ith rotational component (or species) to the anisotropy decay. Theoretically, the sum of all preexponential β_i values must equal the limiting anisotropy value (r_0). The r_0-value is sensitive to fluorescence depolarization due to molecular rotation, nonlinear absorption/emission, ultrafast energy transfer, and instrumentation (e.g., high NA objectives) [66,76]. The rotational time (φ) constant depend on the chemical structure of the fluorophore, the lipid compositions of the bilayer, and how the fluorophore interacts with the bilayer. In some cases, however, the second rotational component is too slow such that the corresponding amplitude is called "residual anisotropy, r_∞" [65]. The initial anisotropy depends on the orientation angle (δ) between the absorbing and emitting dipoles [66,78] such that the maximum theoretical value of r_0 is 0.40 (for one-photon) and 0.57 (for two-photons) [66]. Careful analysis of scattered light, G-factor, and optical depolarization due to the high NA objective [79,80] are critical for meaningful steady-state anisotropy images. The diffusion coefficient (D_R) of a spherical fluorophore depend on its hydrodynamic volume (V) and local viscosity (η) such that $D_R = 1/6\varphi = k_B T/6\eta V$, where k_B and T are the Boltzmann constant and temperature, respectively [66]. In addition, these measurements are also useful for examining the randomized dipole orientation assumption (i.e., $\kappa^2 = 2/3$) for the donor–acceptor pairs in

FRET analysis [70]. It is worth mentioning that in-plane and out-of-plane rotation of lipid and cholesterol analogs can be investigated based on the relative orientation of the molecular dipole orientation in the bilayer with respect to the laser polarization. During the time-resolved anisotropy measurements, however, care must be taken to ensure the vesicle stability during the data acquisition time to eliminate additional depolarization due to vesicle movement.

A typical steady-state, two-photon anisotropy image of DiI-C_{12} in a membrane bleb, isolated from HTB126 epithelial cell, is shown Fig. 4B along with a conventional confocal cross-section image (Fig. 4A). Steady-state anisotropy imaging provides valuable information concerning the degree of order in lipid domains and the dipole-moment orientation distribution of lipid analogs in a lipid bilayer. This anisotropy image was constructed using simultaneous recording of the parallel and perpendicular two-photon fluorescence polarization and the steady-state anisotropy per pixel was calculated using Eq. (3) [33,34,68]. The polarization of the two-photon laser is perpendicular to the microscope axis. These results clearly indicate distinct environmental restrictions and the order of DiI-C_{12} dipole on the membrane. The same approach has been applied to lipid bilayers in GUVs labeled with DiI-C_{12} and Bodipy-PC [34] as well as Bodipy-cholesterol [33].

Additional information on the degree of orientation of the probe in the membrane could be obtained from the angle-dependent fluorescence intensity, $I(\theta)$, around the vesicle contour [75]:

$$I(\theta) = A\left\{\frac{1}{5}+\frac{2}{7}S+\frac{3}{35}P+\frac{6}{7}(S-P)\sin^2(\theta)+P\sin^4(\theta)\right\}, \quad (4)$$

where the order parameters (S and P) that correspond to different moments of the probe orientation distribution and A is a constant reflecting total intensity [75]. These analyses apply to an experimental configuration where the excitation laser is polarized perpendicularly to the microscope axis while the fluorescence emission is detected without polarization analysis. The S order parameter is given by $\langle(3\cos(\delta)-1)/2\rangle$ as compared with $\langle(3-30\cos(\delta)^2+35\cos(\delta)^4/8\rangle$ for the P order parameter [75].

Complementary one-photon anisotropy decays of DiI-C_{12} were also measured in membrane blebs (Fig. 4C) to directly estimate the order parameter (S) as well as the membrane fluidity. Under the same conditions, the anisotropy decays of rhodamine green in buffered solution decays as a single exponential with 180 ± 10 ps at room temperature, which is much shorter than its excited state lifetime (3.95 ± 0.05 ns). As a result, we used rhodamine 6G to estimate the G-factor ($\sim$0.74) using the tail-matching approach [65]. The anisotropy of DiI-C_{12} in a GPMV, isolated from HTB126, decays as a biexponential with rotational time constants of $\varphi_1 = 221 \pm 58$ ps

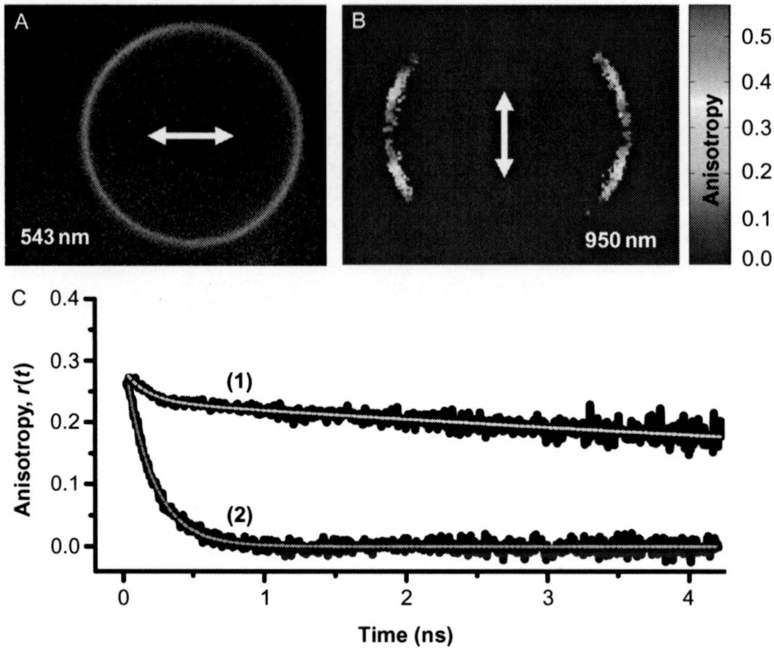

Figure 4 Polarization imaging and time-resolved anisotropy of 2P-fluorescence of DiI-C_{12}-labeled GPMV, isolated from HTB126 epithelial cells. (A) A confocal, cross-sectional image of DiI-C_{12}-labeled membrane bleb was recorded using 543-nm excitation. (B) The corresponding two-photon anisotropy image of DiI-C_{12} in a membrane bleb is also shown using 950-nm excitation. The laser polarization in both confocal and two-photon microscopy is shown here as an arrow within a plane that is perpendicular to the microscope axis. The color code in the anisotropy image represents the average anisotropy value per pixel as calculated using Eq. (2). The polarization of the two-photon laser is perpendicular to the microscope axis and the size of the GUV shown here is ~75 μm. (C) A typical time-resolved, one-photon anisotropy of DiI-C_{12} in a GPMV (curve 1) decays as a biexponential with $\varphi_1 = 221 \pm 58$ ps ($\beta_1 = 0.04 \pm 0.01$) and $\varphi_2 = 24 \pm 9$ ns ($\beta_2 = 0.23 \pm 0.01$). These measurements were carried out using 480-nm laser pulses (repetition rate of 4.2 MHz). As a control, the anisotropy decays of rhodamine green in buffered solution decays as a single exponential with 180 ± 10 ps at room temperature, which is much shorter than its excited state lifetime (3.95 ± 0.05 ns). As a result, we used rhodamine 6G to estimate the *G*-factor (~0.74) using tail-matching approach. (For interpretation of the references to color in this figure legend, the reader is referred to the Web version of this chapter.)

($\beta_1 = 0.04 \pm 0.01$) and $\varphi_2 = 24 \pm 9$ ns ($\beta_2 = 0.23 \pm 0.01$). The initial anisotropy ($r_0 = \beta_1 + \beta_2$) is 0.27 ± 0.02, which is significantly smaller than the theoretical limit (0.4) and may indicate a homo-FRET or another faster dynamic process than our temporal resolution. The slow rotational time can be attributed to the tumbling motion of DiI-C_{12} in the lipid bilayer near the hydrophilic region of the membrane. It is not clear, however, whether the fast rotational time is due to a homo-FRET [81–83] among neighboring DiI-C_{12} molecules or a nonspecific binding of this lipid analog to the membrane. The multiexponential decay of the DiI-C_{12} anisotropy in a membrane bleb (Fig. 4C) suggest a structural/environmental bases to the observed biexponential decay in fluorescence lifetime measurements (Fig. 3D). The order parameter (S) can be calculated using the amplitude fractions of the fast and slow rotational times [34,65], where $S = (r_\infty/r_0)^{0.5} \sim [\beta_2/(\beta_1 + \beta_2)]^{0.5}$, with an estimated $S = 0.92 \pm 0.02$ for DiI-C_{12} in these membrane blebs at room temperature. These results indicate that the order parameter of DiI-C_{12} in GPMVs is larger than that in the lipid bilayer of fluid GUVs (0.48 ± 0.03) [34], which is not surprising due to the presence of elevated cholesterol content [2], sphingolipids, and proteins in the GPMVs. For the same reasons, the lipid bilayers in liquid-disordered GUVs model are more fluidic than those vesicles isolated from the plasma membrane of mammalian cells. Ariola *et al.* have used a similar approach using different lipid and cholesterol analogs for probing the hydrophobic and hydrophilic regions of lipid bilayers in GUVs model [33,34].

5. Translation Diffusion Kinetics of Lipid and Cholesterol Analogs are Sensitive to Phase Domains in Lipid Bilayers

The diffusion of biomolecules in the crowded cellular milieu plays an important role in regulating their activity and function in cell biology. Diffusion also influences how proteins interact during cell signaling as well as the formation of functional biomolecular assemblies. The nature of diffusive motions of cellular proteins may also lead to their segregation into subdomains for functional purposes. The diffusion kinetics of those biomolecules depends on the local cell environment. For example, the two-dimensionally restricted diffusion of membrane-associated proteins enhances their local concentration and, therefore, the probability of protein–protein interactions in a given signaling pathway for regulation purposes. The lipid bilayer thickness, lipid heterogeneities, and lateral pressure in biomembranes may also play a regulatory role of membrane proteins function [61]. Translational diffusion is also a sensitive to lipid phase domains as well as cholesterol content in model membranes [18,31,33,84]. As a result, quantitative and noninvasive studies of diffusion patterns are essential in understanding function of biomolecules and whether these molecules transiently associate with other complexes, lipid rafts [12,13,85], and the cytoskeleton [86].

From the experimental standpoint, lateral diffusion of fluorescent lipid analogs within biomembranes has been reported in different lipid domains using single particle tracking [87,88], fluorescence recovery after photobleaching [89], and FCS [19,31,33,84,90]. FCS relies on fluorescence fluctuations analysis of single-molecules ($10^{-11}-10^{-9}$ M), as they diffuse throughout an open observation volume defined optically [90]. The low concentration of fluorescent labels in FCS experiments is needed to minimize its disturbance of the packing thermodynamics of lipid bilayers or interfere with the biological function of target proteins. The description of a typical FCS experimental setup can be found elsewhere [33,68,90–92]. To separate artifacts from genuine photophysical processes, the FCS system must be routinely calibrated using a photostable fluorophore (e.g., rhodamine 6G or rhodamine green) with a known translational diffusion coefficient ($D_T \sim 2.8 \times 10^{-6}$ cm^2/s) at low laser intensities [90,92,93]. In conventional FCS, the autocorrelation function, $G(\tau)$, of a single species diffusing in three dimensions (3D) is given by [90–92]

$$G_D(\tau) = N^{-1}(1+\tau/\tau_D)^{-1}\left(1+\tau/\omega_0^2\tau_D\right)^{-0.5}, \tag{5}$$

where N is the average number of molecules diffusing in an open 3D Gaussian observation volume with an axial-to-lateral dimension ratio of $\omega_0 = z/\omega_{xy}$. The diffusion time (τ_D), which is the average residence time of a diffusing molecule through an observation volume, is used to calculate the translational diffusion coefficient (D_T). The corresponding autocorrelation function for a diffusing fluorophore in a lipid bilayer (i.e., two-dimensional, 2D) is independent of the structural parameter (ω_0) and is given by [94,95]

$$G_D(\tau) = \frac{1}{N}\frac{1}{(1+\tau/\tau_D)}, \quad \text{where}\ \tau_D = \omega_{xy}^2/4D_T \tag{6}$$

In crowded environments and restrictive domains, molecules are likely to undergo anomalous diffusion, which implies that the mean square displacement of the molecule grows asymptotically as $\langle r^2\rangle = 4\Gamma\tau^{\gamma}$. In this case, the autocorrelation function for such molecular diffusion is given by [96]

$$G_D(\tau) = \frac{1}{N}\frac{1}{[1+(\tau/\tau_D)^{\gamma}]} = \frac{1}{N}\frac{1}{\left[1+\Gamma\tau^{\gamma}/4\omega_{xy}^2\right]}, \tag{7}$$

where Γ (unit: cm^2/s$^{\gamma}$) is the transport coefficient and γ is the degree of anomality (i.e., deviation from Brownian diffusion, where $\gamma = 1$) where $0 < \gamma < 1$. The Stokes–Einstein equation describes the translational diffusion coefficient of a spherical fluorophore in solution (3D) as a function of temperature (T), the Boltzmann constant (k_B), and the hydrodynamic radius (a) of the diffusing species such that $D_T = k_BT/6\pi\eta a$ [66]. The autocorrelation curves are analyzed using OriginPro 7 and the fit goodness was judged by both the residual and χ^2.

Figure 5 (panels A and B) depicts the experimental arrangement for diffusion studies in the bilayer of a vesicle. Keep in mind that the membrane tilting with respect to the microscope axis (or laser polarization) may affect the nature of autocorrelation curve and, therefore, the observed diffusion [96]. A typical autocorrelation curve of DiI-C_{12} fluorescence fluctuation in GPMVs is shown in Fig. 5C. The FCS setup was calibrated using Rh6G (PBS, *p*H 7.4) at room temperature (curve 1), which is best described using Eq. (5) with a diffusion time of $\sim$0.11 ms, which corresponds to a known diffusion coefficient of $D_T = 2.8 \times 10^{-6}$ cm^2/s and a lateral extension of $\sim$0.35 μm for the observation volume. Under the same experimental conditions, the diffusion time of DiI-C_{12} (curve 2) is 0.26 ± 0.01 ms,

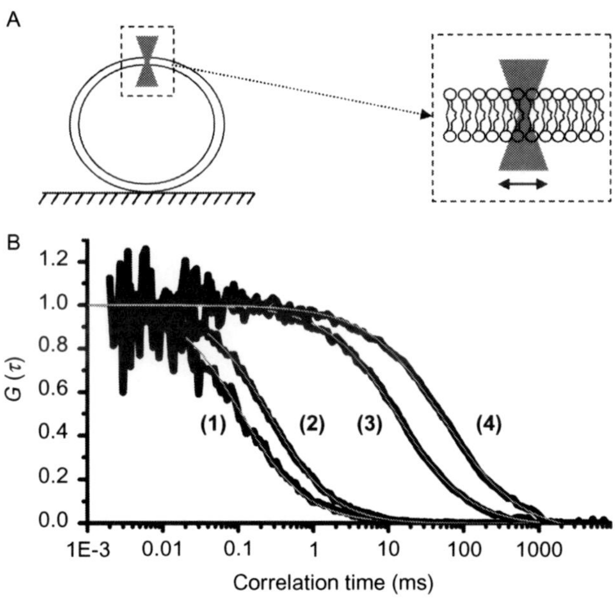

Figure 5 Translational diffusion of DiI-C_{12} in the lipid bilayer of GPMVs that is isolated from HTB126 epithelial cells. The autocorrelation curves of rhodamine 6G (curve 1) and DiI-C_{12} in DMSO (curve 2) are well described by a 3D autocorrelation function with 0.11 and 0.26 ms diffusion time constant, respectively. In the lipid bilayer of a GPMV (curve 3), however, DiI-C_{12} diffuses much slower ($\tau_D \sim 13$ ms) than in solution. In the plasma membrane of HTB126 epithelial cell, the diffusion time of DiI-C_{12} is 55 ± 10 ms (curve 4). The anomalous diffusion of DiI-C_{12} in the lipid bilayer (either in GPMVs or the plasma membrane) is described satisfactorily with 2D autocorrelation function with $\gamma = 0.8 \pm 0.1$. These autocorrelation curves were recorded using 543-nm laser (cw), 1.4 *NA* objective (oil immersion), and an optical fiber of 50-μm diameter as a confocal pinhole.

which yields $D_T \sim 1.2 \times 10^{-6}$ cm^2/s in DMSO at room temperature. In GPMVs experiment, the HTB126 cells were treated with GPMV buffer for one hour prior to the experiment to induce membrane blebing. In the membrane of a GPMV (curve 3), however, DiI-C_{12} subdiffuses at much slower time scale ($\tau_D = 13 \pm 5$ ms; $n = 9$ vesicles) than in solution with anomality coefficient $\gamma = 0.89 \pm 0.03$. The estimated diffusion coefficient of DiI-C_{12} in these membrane blebs is $(2 \pm 1) \times 10^{-8}$ cm^2/s. The average number of molecules in the observation volume is 5 ± 2 during these measurements. Alternatively, the autocorrelation curve can be described using a two diffusing species model with a fast diffusion time of 5 ± 1 ms with an amplitude fraction of $\sim$0.30. In the plasma membrane of intact HTB126 epithelial cell, without chemical treatment for blebing, the anomalous diffusion time of DiI-C_{12} is 61 ± 20 ms (curve 4) with $\gamma = 0.86 \pm 0.16$. The corresponding diffusion coefficient of DiI-C_{12} in intact plasma membrane, under resting conditions, is $(4.3 \pm 0.9) \times 10^{-9}$ cm^2/s. The observed enhancement of DiI-C_{12} diffusion in membrane blebs can be attributed to the presence of the membrane proteins in the plasma membrane of intact cells. These results are in agreement with the notion that membrane blebing has interrupted the plasma membrane interactions with the cytoskeleton. It is not clear, however, whether the exoskeletal protein matrix has been retained in the chemically induced GPMVs or not. Similar conclusions have been reported previously on muscle cells using fluorescence photobleaching recovery [97].

The same experimental approach has been used recently to investigate the diffusion of a new Bodipy-cholesterol derivative in the bilayer of ternary GUVs [33]. In these studies, translational and rotational diffusion measurements were combined using FCS and time-resolved anisotropy, respectively, and it was concluded that Bodipy-cholesterol derivative is likely to diffuse as a cluster with a few other lipid molecules in the bilayer of GUVs. In addition, the diffusion of Bodipy-cholesterol can be described well using Stokes–Einstein models as compared with the hydrodynamic model by Saffman and Delbrück [98]. In the hydrodynamic model by Saffman and Delbrück [98], the translational diffusion coefficient of a transmembrane protein in a lipid bilayer depends on the viscosity (of both membrane, η_m, and surrounding solvent, η_w), membrane thickness (h), and protein radius (a) such that

$$D_T = \frac{k_B T}{4\pi\eta_m h}[\mathrm{In}(\eta_m h/\eta_w a) - 0.5772] \tag{8}$$

The model also provides a theoretical prediction for the translational-to-rotational diffusion coefficient ratio (D_T/D_R), which depends quadratically on the hydrodynamic radius of the diffusing protein in a lipid bilayer [98].

$$D_T/D_R = [\text{In}(\eta_m h/\eta_w a) - 0.5772]a^2 \qquad (9)$$

As a result, using these multiscale diffusion (rotational and translational) measurements and the corresponding temporal-to-spatial scaling, one would expect that the rotational diffusion of a molecule spans shorter (or microscopic) distances when compared with the long-range (or macroscopic) translational diffusion [87]. This spatiotemporal scaling may also be important in differentiating between Brownian and non-Brownian (or anomalous) diffusion in biological membranes due to molecular crowding, transmembrane proteins, and cytoskeletal confinement. Recent studies on various peptides and transmembrane proteins in 1-stearoyl-2-oleoyl-*sn*-glycero-3-phosphocholine (SOPC) GUVs have suggested that the measured translational diffusion coefficient disagrees with the values predicted using the Saffman–Delbrück model [99]. The authors also indicate that their experimental results agree with a heuristic Stokes–Einstein-like expression that is weighted by a characteristic length for dimensional reasons. Mesoscopic simulation studies have also been reported to quantify the size-dependent diffusion properties of membrane inclusions [100]. It was suggested that the Saffman–Delbrück model describe well the translational diffusion of molecules with a smaller radius than the critical value of $h\eta_m/2\eta_w \sim 7.4$ nm. However, the authors reported significant deviations from the hydrodynamic model for molecules with a radius larger than the critical value [100]. In contrast to the lateral diffusion, the authors also reported that the size dependence of the rotational diffusion coefficient (D_R) follows the predicted hydrodynamic scaling over the entire size range studied.

6. Conclusion and Outlook

A considerable number of studies over the last decade have focused on the structural and dynamical aspects of lateral segregation of lipid domains in model membranes. While model membranes are not an exact mimic of the complex biological membranes, they are ideal for fundamental research in lipid bilayer assembly under controlled environments. From these model membranes, the underlying physical principles that govern the lateral membrane heterogeneity and lipid-domain formation can be understood. Researchers have used these membrane models to develop new experimental approaches prior to their applications in cellular membranes. In addition, model membranes are ideal systems to test theories associated with different aspects of membrane assemblies and biophysics. It is noteworthy that the phase separation in model membranes may not be absolute and, therefore, lipid exchange is likely to exist, which would also imply some degree of partitioning of lipid analogs in different phases. Membrane blebs (or GPMVs) bring researchers a step closer to mimicking biological membranes with the inherent variety of lipids and proteins. The creation of these GPMVs, however, requires chemically-induced apoptosis whose effect on the membrane integrity remains unclear. The lack of actin-based cortical cytoskeleton in GPMVs, which is believed to regulate lipid heterogeneity, is likely to influence the structure–function relationship in biological membrane. Accordingly, care must also be taken to avoid overreaching in GPMV data interpretation and their relevance to the structure and function of biological membranes. The relevance of temperature-dependence of phase segregation in GPMV to the plasma membrane in living cells requires further investigation.

These model membrane studies serve as a valuable point of reference, but now it is time to shift gears toward understanding of complex biological membranes, under physiological conditions and cell activation, as compared with models systems. These comparative studies will allow researchers to elucidate the role of cytoskeleton and protein polymerization in the lateral heterogeneity of plasma membrane, which is not fully understood. With the advances in sophisticated imaging technologies and stage-incubators, new opportunities are emerging that will help quantify the dynamics of lipid-domain formation as a function of cholesterol content and the presence of membrane protein in more realistic model systems. In addition, it is likely that the diffusion of biomolecules on the membrane is dependent on the time (i.e., anomalous diffusion) associated with a given experimental design due to molecular crowding and restrictions enforced by the cytoskeleton-induced domains.

The inherent biocomplexity of biomembranes calls for a multiparametric fluorescence approach as described here, which integrates quantitative, noninvasive micro-spectroscopy techniques for biological and biomimetic membrane studies with molecular sensitivity. The projected wealth of information from this experimental approach is critical for model building and theory testing. In addition to conventional confocal and DIC imaging for visualization and time lapses studies, fluorescence lifetime imaging is sensitive to lateral heterogeneity in bilayers, intermolecular interactions, and the chemical structure of the fluorescent labels. Based on their chemical structures, the lipid and cholesterol analogs interact with lipid bilayers differently. For example, some analogs favor the hydrophobic core as compared to others with high affinity for the hydrophilic region of the membrane. Polarization-analyzed fluorescence imaging (in space and/or time) is the technique of choice for probing lipid phases, order parameters, membrane fluidity, and how the fluorescent analogs interact with the lipid bilayer. Lipids and cholesterol also undergo translational diffusion and intermolecular association with other bilayer constituents, which could be "transient" or "long-lived". Different modalities of FCS represent an ideal method for diffusion and chemical kinetics studies in lipid bilayers with single-molecule sensitivity. The nature of diffusion processes (i.e., Brownian vs. non-Brownian or anomalous diffusion) can also be investigated using FCS within the context of the crowded, heterogeneous structure of the plasma membrane in living cells or the ternary phase domains in model membranes. From the sample perspective, however, care must be taken to control different thermodynamic parameters that may influence the biophysical observable under investigation. Only then, one could elucidate the molecular underpinning of structure–function relationship in biological membranes within the context of cell biology. Combining these integrated fluorescence methods with nanofabrication technology is a step forward this goal. These fabricated surfaces allow for site-specific manipulation of different aspect of biomembranes and cell signaling.

The multiscale fluorescence approach is likely to help capturing some of the biophysical cues of the elusive membrane domains in living cells. However, we may reconsider the notion of lipid domains "or rafts" not as a long-lived stable entity (as in model membranes) but rather as a transient, dynamic association of membrane protein(s) [4] with specific types of lipids and in the presence of cholesterol. Such transient assembly will be formed in response to extracellular or intracellular cues to execute a biological function. This response will lead to structural changes of a target membrane receptor followed by rearrangement of neighboring lipids for minimizing the line tension and associated free energy. The lifetime of the newly formed transient protein–lipid assembly will be limited to the time needed for relaying the biological cue that are structurally encoded to the next participant protein in a signaling cascade. The numerous types of lipids and the presence of membrane proteins in the plasma membrane of mammalian cells are likely to regulate this notion of transient formation for these functional assemblies, particularly under physiological conditions of temperature and molecular crowding. At the heart of this view are the inherent multiscale processes associated with the formation of functional domains, whether it is tumbling motion or translational diffusion of lipids, diffusion and structural changes of membrane proteins, intermolecular interactions, receptor-activated signaling cascade, and the local-to-global cellular response.

ACKNOWLEDGMENTS

I thank my former students (Deepti Mudaliar, Florly Ariola, and Christine Cornejo) for their help in obtaining some of the results used in this chapter. I also thank Ashli Tucker (Department of Chemistry and Biochemistry, University of Minnesota Duluth) and

Dr. Erin Sheets (Department of Pharmacy Practice and Pharmaceutical Sciences, College of Pharmacy, University of Minnesota Duluth) for their valuable editorial comments. This work was supported in part by NSF (MCB0718741) and NIH (AG030949). Additional support was also provided by the office of the Vice President for Research (Grant-In-Aid for Research, Artistry, and Scholarship), University of Minnesota, the Swenson College of Science and Engineering and the College of Pharmacy at the University of Minnesota Duluth.

REFERENCES

[1] M. Edidin, Lipids on the frontier: a century of cell-membrane bilayers, Nat. Rev. Mol. Cell Biol. 4 (2003) 415.
[2] R.B. Gennis, Biomembranes: molecular structure and function, Springer-Verlag, New York (1989).
[3] V. Kiessling, C. Wan, L.K. Tamm, Domain coupling in asymmetric lipid bilayers, Biochim. Biophys. Acta 1788 (2009) 64–71.
[4] M. Edidin, The state of lipid rafts: from model membranes to cells, Annu. Rev. Biophys. Biomolec. Struct. 32 (2003) 257–283.
[5] K.A. Field, D. Holowka, B. Baird, Fc(epsilon)RI-mediated recruitment of p53/56lyn to detergent-resistant membrane domains accompanies cellular signaling, Proc. Natl. Acad. Sci. USA 92 (1995) 9201–9205.
[6] A. Pralle, P. Keller, E.-L. Florin, K. Simons, J.K.H. Hörber, Sphingolipid–cholesterol rafts diffuse as small entities in the plasma membrane of mammalian cells, J. Cell Biol. 148 (2000) 997–1007.
[7] M. Simpson-Holley, D. Ellis, D. Fisher, D. Elton, J. McCauley, P. Digard, A functional link between the actin cytoskeleton and lipid rafts during budding of filamentous influenza virions, Virology 301 (2002) 212.
[8] L. Rajendran, K. Simons, Lipid rafts and membrane dynamics, J. Cell Sci. 118 (2005) 1099–1102.
[9] S. Munro, Lipid rafts: elusive or illusive? Cell 115 (2003) 377.
[10] S.L. Veatch, P. Cicuta, P. Sengupta, A. Honerkamp-Smith, D. Holowka, B. Baird, Critical fluctuations in plasma membrane vesicles, ACS Chem. Biol. 3 (2008) 287–293.
[11] D. Holowka, Biomembrane liquid–liquid phase separation and detergent resistance: a relationship strengthened, Biochem. J. 424 (2009) e5–e6.
[12] K. Simons, Lipid rafts and signal transduction, Nat. Rev. Mol. Cell Biol. 1 (2000) 31–39.
[13] K. Simons, E. Ikonen, Functional rafts in cell membranes, Nature 387 (1997) 569–572.
[14] Y.H. Chan, S.G. Boxer, Model membrane systems and their applications, Curr. Opin. Chem. Biol. 11 (2007) 581–587.
[15] C.H. Spink, M.D. Yeager, G.W. Feigenson, Partitioning behavior of indocarbocyanine probes between coexisting gel and fluid phases in model membranes, Biochim. Biophys. Acta. Biomembr. 1023 (1990) 25–33.
[16] R.E. Brown, Sphingolipid organization in biomembrane: what physical studies of model membranes reveal, J. Cell Sci. 111 (1998) 1–9.
[17] E. London, How principles of domain formation in model membranes may explain ambiguities concerning lipid raft formation in cells, Biochim. Biophys. Acta 1746 (2005) 203–220.
[18] N. Kahya, D. Scherfeld, K. Bacia, B. Poolman, P. Schwille, Probing lipid mobility of raft-exhibiting model membranes by fluorescence correlation spectroscopy, J. Biol. Chem. 278 (2003) 28109–28115.

[19] P. Schwille, J. Korlach, W.W. Webb, Fluorescence correlation spectroscopy with single-molecule sensitivity on cell and model membranes, Cytometry 36 (1999) 176–182.
[20] C. Dietrich, L.A. Bagatolli, Z.N. Volovyk, N.L. Thompson, M. Levi, K. Jacobson, E. Gratton, Lipid rafts reconstituted in model membranes, Biophys. J. 80 (2001) 1417–1428.
[21] S.N. Ahmed, D.A. Brown, E. London, On the origin of sphingolipid/cholesterol-rich detergent-insoluble cell membranes: Physiological concentrations of cholesterol and sphingolipid induce formation of a detergent-insoluble, liquid-ordered lipid phase in model membranes, Biochemistry 36 (1997) 10944–10953.
[22] C. Schwieger, A. Blume, Interaction of poly(L-arginine) with negatively charged DPPG membranes: calorimetric and monolayer studies, Biomacromolecules 10 (2009) 2152–2161.
[23] M. Arcisio-Miranda, M.P. dos Santos Cabrera, K. Konno, M. Rangel, J. Procopio, Effects of the cationic antimicrobial peptide eumenitin from the venom of solitary wasp Eumenes rubronotatus in planar lipid bilayers: surface charge and pore formation activity, Toxicon 51 (2008) 736–745.
[24] O.Y. Shatursky, T.M. Volkova, N.H. Himmelreich, E.V. Grishin, The geometry of the ionic channel lumen formed by alpha-latroinsectotoxin from black widow spider venom in the bilayer lipid membranes, Biochim. Biophys. Acta 1768 (2007) 2757–2763.
[25] V.V. Teplova, A.A. Tonshin, P.A. Grigoriev, N.E. Saris, M.S. Salkinoja-Salonen, Bafilomycin A1 is a potassium ionophore that impairs mitochondrial functions, J. Bioenerg. Biomembr. 39 (2007) 321–329.
[26] Olga Wesołowska, Krystyna Michalak, Jadwiga Maniewska, A.B. Hendrich, Giant unilamellar vesicles—a perfect tool to visualize phase separation and lipid rafts in model systems, Acta Biochem. Pol. 56 (2009) 33–39.
[27] E. Farge, P.F. Devaux, Shape changes of giant liposomes induced by an asymmetric transmembrane distribution of phospholipids, Biophys. J. 61 (1992) 347–357.
[28] A. Papadopulos, S. Vehring, I. Lopez-Montero, L. Kutschenko, M. Stockl, P.F. Devaux, M. Kozlov, T. Pomorski, A. Herrmann, Flippase activity detected with unlabeled lipids by shape changes of giant unilamellar vesicles, J. Biol. Chem. 282 (2007) 15559–15568.
[29] G. Staneva, M.I. Angelova, K. Koumanov, Phospholipase A2 promotes raft budding and fission from giant liposomes, Chem. Phys. Lipids 129 (2004) 53–62.
[30] E. Ambroggio, F. Separovic, B. John, G.D. Fidelio, L.A. Bagatolli, Direct visualization of membrane leakage induced by the antibiotic peptides: Maculatin, Citropin and Aurein, Biophys. J. 89 (2005) 1874–1881.
[31] J. Korlach, P. Schwille, W.W. Webb, G.W. Feigenson, Characterization of lipid bilayer phases by confocal microscopy and fluorescence correlation spectroscopy, Proc. Natl. Acad. Sci. USA 96 (1999) 8461–8466.
[32] T. Baumgart, S.T. Hess, W.W. Webb, Imaging coexisting fluid domains in biomembrane models coupling curvature and line tension, Nature 425 (2003) 821–824.
[33] F.S. Ariola, Z. Li, C. Cornejo, R. Bittman, A.A. Heikal, Membrane fluidity and lipid order in ternary giant unilamellar vesicles using a new Bodipy-cholesterol derivative, Biophys. J. 96 (2009) 2696–2708.
[34] F.S. Ariola, D.J. Mudaliar, R.P. Walvick, A.A. Heikal, Dynamics imaging of lipid phases and lipid-marker interactions in model biomembranes, Phys. Chem. Chem. Phys. 39 (2006) 4517–4529.
[35] L.A. Bagatolli, E. Gratton, Two photon fluorescence microscopy of coexisting lipid domains in giant unilamellar vesicles of binary phospholipid mixtures, Biophys. J. 78 (2000) 290–305.

[36] C. Leidy, W.F. Wolkers, K. Jorgensen, O.G. Mouritsen, J.H. Crowe, Lateral organization and domain formation in a two-component lipid membrane system, Biophys. J. 80 (2001) 1819–1828.
[37] T.-Y. Wang, R. Leventis, J.R. Silvius, Fluorescence-based evaluation of the partitioning of lipids and lipidated peptides into liquid-ordered lipid microdomains: a model for molecular partitioning into "lipid rafts" Biophys. J. 79 (2000) 919–933.
[38] G.W. Feigenson, Phase boundaries and biological membranes, Annu. Rev. Biophys. Biomol. Struct. 36 (2007) 63–77.
[39] L.A. Bagatolli, Membranes and fluorescence microscopy, in: C.D. Geddes, J.R. Lakowicz (Eds.), Annual Review in Fluorescence 2007, Springer, Heidelberg, 2007, pp. 33–51.
[40] Y. Li, L.K. Tamm, Structure and plasticity of the human immunodeficiency virus gp41 fusion domain in lipid micelles and bilayers, Biophys. J. 93 (2007) 876–885.
[41] A. Hinderliter, R.L. Biltonen, P.F.F. Almeida, Lipid modulation of protein-induced membrane domains as a mechanism for control of signal transduction, Biochemistry 43 (2004) 7102–7110.
[42] N. Kahya, D.A. Brown, P. Schwille, Raft partitioning and dynamic behavior of human placental alkaline phosphatase in giant unilamellar vesicles, Biochemistry 44 (2005) 7479–7489.
[43] N. Kahya, D.A. Wiersma, B. Poolman, D. Hoekstra, Spatial organization of Bacteriorhodopsin in model membranes, J. Biol. Chem. 277 (2002) 39304–39311.
[44] N. Kahya, Protein–protein and protein–lipid interactions in domain-assembly: lessons from giant unilamellar vesicles, Biochim. Biophys. Acta 1798 (2010) 1392–1398.
[45] M.K. Doeven, J.H.A. Folgering, V. Krasnikov, E.R. Geertsma, G. van den Bogaart, B. Poolman, Distribution, lateral mobility and function of membrane proteins incorporated into giant unilamellar vesicles, Biophys. J. 88 (2005) 1134–1142.
[46] P. Girard, J. Pécréaux, G. Lenoir, P. Falson, J.-L. Rigaud, P. Bassereau, A new method for the reconstitution of membrane proteins into giant unilamellar vesicles, Biophys. J. 87 (2004) 419–429.
[47] C.H. Arnaud, Simulating life's envelopes: models provide clues about lipid behavior in cell membrane, but they may have reached their limits, Chem. Eng. News 87 (2009) 31–33.
[48] R.E. Scott, Plasma membrane vesiculation: a new technique for isolation of plasma membranes, Science 194 (1976) 743–745.
[49] T. Baumgart, A.T. Hammond, P. Sengupta, S.T. Hess, D.A. Holowka, B.A. Baird, W.W. Webb, Large scale fluid/fluid phase separation of proteins and lipids in giant plasma membrane vesicles, Proc. Natl. Acad. Sci. USA 104 (2007) 3165–3170.
[50] P. Sengupta, A. Hammond, D. Holowka, B. Baird, Structural determinants for partitioning of lipids and proteins between coexisting fluid phases in giant plasma membrane vesicles, Biochim. Biophys. Acta 1778 (2008) 20–32.
[51] X. Han, N.L. Smith, D. Sil, D.A. Holowka, F.W. McLafferty, B.A. Baird, IgE receptor-mediated alteration of membrane–cytoskeleton interactions revealed by mass spectrometric analysis of detergent-resistant membranes, Biochemistry 48 (2009) 6540–6550.
[52] I. Levental, F.J. Byfield, P. Chowdhury, F. Gai, T. Baumgart, P.A. Janmey, Cholesterol-dependent phase separation in cell-derived giant plasma membrane vesicles, Biochem. J. 424 (2009) 163–167.
[53] S.L. Veatch, S.L. Keller, Organization in lipid membranes containing cholesterol, Phys. Rev. Lett. 89 (2002) 268101–268104.
[54] O. Garvik, P. Benediktson, A.C. Simonsen, J.H. Ipsen, D. Wüstner, The fluorescent cholesterol analog dehydroergosterol induces liquid-ordered domains in model membranes, Chemistry and Physics of Lipids 159 (2009) 114–118.

[55] W. Denk, J.H. Strickler, W.W. Webb, Two-photon laser scanning fluorescence microscopy, Science 248 (1990) 73–76.
[56] W.R. Zipfel, R.M. Williams, W.W. Webb, Nonlinear magic: multiphoton microscopy in the biosciences, Nat. Biotechnol. 21 (2003) 1369–1377.
[57] P.T.C. So, C.Y. Dong, B.R. Masters, K.M. Berland, Two-photon excitation fluorescence microscopy, Annual Review of Biomedical Engineering 2 (2000) 399–429.
[58] L.A. Bagatolli, E. Gratton, A correlation between lipid domain shape and binary phospholipid mixture composition in free standing bilayers: a two-photon fluorescence microscopy study, Biophys. J. 79 (2000) 434–447.
[59] K. Gaus, E. Gratton, E.P.W. Kable, A.S. Jones, I. Gelissen, L. Kritharides, W. Jessup, Visualizing lipid structure and raft domains in living cells with two-photon microscopy, Proc. Natl. Acad. Sci. USA 100 (2003) 15554–15559.
[60] L. Jin, A.C. Millard, J.P. Wuskell, H.A. Clark, L.M. Loew, Cholesterol-enriched lipid domains can be visualized by di-4-ANEPPDHQ with linear and nonlinear Optics, Biophys. J. 89 (2005) L04–L06.
[61] D. Marsh, Protein modulation of lipids, and *vice-versa*, in membranes, Biochim. Biophys. Acta 1778 (2008) 1545–1575.
[62] A. Periasamy, R. M. Clegg, FLIM microscopy in biology and medicine, Chapman & Hall, CRC Press, Taylor and Francis Group, New York (2010).
[63] W. Becker, Advanced time-correlated single-photon counting techniques (2005) 1–376.
[64] D.V. O'Connor, D. Phillips, Time-Correlated Single Photon Counting, Academic Press, London, 1984.
[65] J.R. Lakowicz, Principles of Fluorescence Spectroscopy, Springer, Singapore, 2006.
[66] J. Lakowicz, Principles of Fluorescence Spectroscopy, Kluwer Academic/Plenum Publishers, New York, 1999.
[67] A.M. Davey, R.P. Walvick, Y. Liu, A.A. Heikal, E.D. Sheets, Membrane order and molecular dynamics associated with IgE receptor cross-linking in mast cells, Biophys. J. 92 (2007) 343–355.
[68] Q. Yu, M. Proia, A.A. Heikal, Integrated biophotonics approach for non-invasive, multiscale studies of biomolecular and cellular biophysics, J. Biomed. Opt. 13 (2008) 041315.
[69] L.M.S. Loura, A. Fedorov, M. Prieto, Partition of membrane probes in a gel/fluid two-component lipid system: a fluorescence resonance energy transfer study, Biochim. Biophys. Acta 1467 (2000) 101–112.
[70] S.S. Vogel, C. Thaler, S.V. Koushik, Fanciful FRET, Sci. STKE 2006 (2006) re2.
[71] L.M.S. Loura, A. Fedorov, M. Prieto, Fluid-fluid membrane microheterogeneity: a fluorescence resonance energy transfer study, Biophys. J. 80 (2001) 776–788.
[72] D. Alvarez, P. Medina, M. Moscoso, Fluorescence lifetime imaging from time resolved measurements using a shape-based approach, Opt. Express. 17 (2009) 8843–8855.
[73] M.A. Digman, V.R. Caiolfa, M. Zamai, E. Gratton, The phasor approach to fluorescence lifetime imaging analysis, Biophys. J. 94 (2008) L14–L16.
[74] R.A. Colyer, C. Lee, E. Gratton, A novel fluorescence lifetime imaging system that optimizes photon efficiency, Microsc. Res. Tech. 71 (2008) 201–213.
[75] C.K. Haluska, A.P. Schröder, P. Didier, D. Heissler, G. Duportail, Y. Mély, C.M. Marques, Combining fluorescence lifetime and polarization microscopy to discriminate phase separated domains in giant unilamellar vesicles, Biophys. J. 95 (2008) 5737–5747.
[76] S.S. Vogel, C. Thaler, P.S. Blank, S.V. Koushik, Time resolved fluorescence anisotropy, Taylor & Francis, Boca Raton, 2010.

[77] L. Davenport, P. Targowski, Submicrosecond phospholipid dynamics using a long-lived fluorescence emission anisotropy probe, Biophys. J. 71 (1996) 1837–1852.
[78] A. Volkmer, V. Subramaniam, D.J.S. Birch, T.M. Jovin, One- and two-photon excited fluorescence lifetimes and anisotropy decays of green fluorescent proteins, Biophys. J. 78 (2000) 1589–1598.
[79] D. Axelrod, Carbocyanine dye orientation in red-cell membrane studied by microscopic fluorescence polarization, Biophys. J. 26 (1979) 557–573.
[80] D. Axelrod, Fluorescence polarization microscopy, Methods Cell Biol. 30 (1989) 333–352.
[81] M. Rao, S. Mayor, Use of Förster's resonance energy transfer microscopy to study lipid rafts, Biochim. Biophys. Acta 1746 (2005) 221–233.
[82] S. Scolari, S. Engel, N. Krebs, A.P. Plazzo, R.F.M. De Almeida, M. Prieto, M. Veit, A. Herrmann, Lateral distribution of the transmembrane domain of influenza virus Hemagglutinin revealed by time-resolved fluorescence imaging, J. Biol. Chem. 284 (2009) 15708–15716.
[83] I. Gautier, M. Tramier, C. Durieux, J. Coppey, R.B. Pansu, J.C. Nicolas, K. Kemnitz, M. Coppey-Moisan, Homo-FRET microscopy in living cells to measure monomer-dimer transition of GFP-tagged proteins, Biophys. J. 80 (2001) 3000–3008.
[84] J. Korlach, T. Baumgart, W.W. Webb, G.W. Feigenson, Detection of motional heterogeneities in lipid bilayer membranes by dual probe fluorescence correlation spectroscopy, Biochim. Biophys. Acta Biomembr. 1668 (2005) 158.
[85] K. Simons, W.L.C. Vaz, Model systems, lipid rafts, and cell membranes, Annu. Rev. Biophys. Biomol. Struct. 33 (2004) 269–295.
[86] N. Yamaguchi, M.N. Fukuda, Golgi retention mechanism of beta-1,4-galactosyltransferase. Membrane-spanning domain-dependent homodimerization and association with alpha- and beta-tubulins, Biol. Chem. 270 (1995) 12170–12176.
[87] M.J. Saxton, K. Jacobson, Single-particle tracking: applications to membrane dynamics, Annu. Rev. Biophys. Biomol. Struct. 26 (1997) 373–399.
[88] E.D. Sheets, G.M. Lee, R. Simson, K. Jacobson, Transient confinement of a glycosylphosphatidylinositol-anchored protein in the plasma membrane, Biochemistry 36 (1997) 12449–12458.
[89] C. Klein, T. Pillot, J. Chambaz, B. Drouet, Determination of plasma membrane fluidity with a fluorescent analogue of sphingomyelin by FRAP measurement using a standard confocal microscope, Brain Res. Protoc. 11 (2003) 46.
[90] S.T. Hess, S. Huang, A.A. Heikal, W.W. Webb, Biological and chemical applications of fluorescence correlation spectroscopy: a review, Biochemistry 41 (2002) 697–705.
[91] U. Haupts, S. Maiti, P. Schwille, W.W. Webb, Dynamics of fluorescence fluctuations in green fluorescent protein observed by fluorescence correlation spectroscopy, Proc. Natl. Acad. Sci. USA 95 (1998) 13573–13578.
[92] A.A. Heikal, S.T. Hess, G.S. Baird, R.Y. Tsien, W.W. Webb, Molecular spectroscopy and dynamics of intrinsically fluorescent proteins: Coral red (dsRed) and yellow (Citrine), Proc. Natl. Acad. Sci. USA 97 (2000) 11996–12001.
[93] N.V. Visser, M.A. Hink, A.v. Hoek, A.J.W.G. Visser, Comparison between fluorescence correlation spectroscopy and time-resolved fluorescence anisotropy as illustrated with a fluorescent dextran conjugate, J. Fluorescence 9 (1999) 251–255.
[94] J. Ries, P. Schwille, New concepts for fluorescence correlation spectroscopy on membranes, Phys. Chem. Chem. Phys. 10 (2008) 3487–3497.
[95] P. Sengupta, J. Balaji, S. Maiti, Measuring diffusion in cell membranes by fluorescence correlation spectroscopy, Methods (San Diego, CA, United States) 27 (2002) 374–387.
[96] M. Weiss, H. Hashimoto, T. Nilsson, Anomalous protein diffusion in living cells as seen by fluorescence correlation spectroscopy, Biophys. J. 84 (2003) 4043–4052.

[97] D.W. Tank, E.-S. Wu, W.W. Webb, Enhanced molecular diffusibility in muscle membrane blebs: Release of lateral constraints, J. Cell Biol. 92 (1982) 207–212.
[98] P.G. Saffman, M. Delbrück, Brownian motion in biological membranes, Proc. Natl. Acad. Sci. USA 72 (1975) 3111–3113.
[99] Y. Gambin, R. Lopez-Esparza, M. Reffay, E. Sierecki, N.S. Gov, M. Genest, R.S. Hodges, W. Urbach, Lateral mobility of proteins in liquid membranes revisited, Proc. Natl. Acad. Sci. USA 103 (2006) 2098–2102.
[100] G. Guigas, M. Weiss, Size-dependent diffusion of membrane inclusions, Biophys. J. 91 (2006) 2393–2398.

CHAPTER SIX

Electromechanical Basis for the Interaction Between Osteoblasts and Negatively Charged Titanium Surface

Ekaterina Gongadze,[1] Šarka Perutková,[2] Veronika Kralj-Iglič,[3] Ursula van Rienen,[1] Ulrich Beck,[4] Aleš Iglič,[2] *and* Doron Kabaso[2,*]

Contents

Abstract

Nonetheless, titanium (Ti) surface and osteoblasts are negatively charged, there is an attractive interaction between them, which we aim to explain here theoretically. It is shown that adhesion of positively charged proteins with internal charge distribution may give rise to initial attractive interactions between the Ti surface and the osteoblast membrane. A dynamic model of the osteoblast attachment is presented in order to study the impact of geometrically structured Ti surfaces on the osteoblast attachment. It is indicated that

* Corresponding author: Tel.: +386 1 476 88 25; Fax: +386 1 476 88 50.
E-mail address: doron.kabaso@fe.uni-lj.si

1 Institute of General Electrical Engineering, University of Rostock, Rostock, Germany
2 Laboratory of Biophysics, Faculty of Electrical Engineering, University of Ljubljana, Ljubljana, Slovenia
3 Laboratory of Clinical Biophysics, Faculty of Medicine, University of Ljubljana, Ljubljana, Slovenia
4 Institute for Electronic Appliances and Circuits, University of Rostock, Germany

Advances in Planar Lipid Bilayers and Liposomes, Volume 13
ISSN 1554-4516, DOI: 10.1016/B978-0-12-387721-5.00006-7

membrane-bound protein complexes (PCs) may increase the membrane protrusion growth between the osteoblast and the grooves on Ti surface and thereby facilitate the attachment of osteoblasts to the Ti surface. On the other hand, a strong local adhesion due to electrostatic forces may locally trap the osteoblast membrane and hinder the further spreading of the osteointegration boundary.

1. Role of Positively Charged Macroions in the Interaction Between an Implant Surface and Osteoblastic Cells

The interactions between implant and surrounding bone tissue are essential for the successful integration of a bone implant. Due to its good biocompatibility, titanium (Ti) and its alloys [1] are widely used in a variety of orthopedic and dental implants. The oxide layer, naturally formed on it, is thought to be mainly responsible for its good biocompatible behavior [2]. The nearly free electrons and the dissolution of metal cations [3] partially lead to a negative surface potential, negative surface charge density of Ti implants. In addition, the negative surface potential is a consequence of a preferential adsorption of chloride and hydroxyl anions, incorporated in the structure of the TiO_2 layer [2,4]. We should note that a contribution to the negative surface potential would be introduced by an applied voltage to the electrostimulative implants [3]. The negatively charged surface of the Ti implant attracts cations and repels anions, and consequently an electric double layer is formed [1,2]. Therefore, the surface potential strongly depends on the surrounding ionic strength [2].

The adhesion of osteoblasts to the Ti surface is a prerequisite for the successful osteointegration of an implant [4]. It was suggested that the contact between the cell membrane of osteoblasts and the Ti oxide surface is established in two steps [4,5]. Firstly, the osteoblast cell membrane makes a nonspecific contact due to the electrostatics, followed by a second step, where the specific binding takes place. Many studies in the past have shown that the negative surface potential of Ti implants promotes the osteoblast adhesion and consequently the new bone formation [6,7]. Since osteoblasts are negatively charged [6], they would be electrostatically repelled by the negatively charged Ti surface if some other attractive forces would not be present in the system. Therefore, recently a mechanism of osteoblast binding to the implant surface was proposed, assuming that positively charged proteins, attached to the negatively charged implant surface, serve as a substrate for the subsequent attachment of the osteoblasts [6,8,9].

The present chapter investigates the physical mechanisms responsible for the charged protein-mediated attractive interactions between the negatively charged osteoblast membrane and the negatively charged Ti implant surfaces.

Additionally, contributions of curvature-inducing protein complexes (PCs) to the membrane free energy may explain puzzling experimental data, demonstrating no osteointegration between narrow grooves and a considerable osteointegration between the wide grooves of Ti surface implants [10–12].

2. On the Origin of the Initial Attraction Between can Osteoblast and a Ti Implant Surface

We assume that the initial attractive interaction between the negatively charged Ti surface and the negatively charged osteoblast membrane surface is mediated by adhesion of positive macroions (proteins) with an internal charge distribution [8]. This adhesion originates in Coulomb interaction between the negatively charged surface and positively charged proteins. In our theoretical model, the single spheroidal protein is composed of two positive charges (each of the valency Z_s), separated by a distance D equal to the length of the protein along its axis.

To predict the orientation of the proteins near a charged implant surface, we performed Monte Carlo (MC) simulations of the distribution and the orientation of charged spherical proteins in the close vicinity of the charged planar surface with the surface charge density σ_{eff}. The simulation system consisted of a box with a volume LY^2, where $L = 10$ nm is the distance between these two surfaces and Y^2 is the area of both planar surfaces that confine the system. We assume that electric field changes just along x-axis perpendicular to the charged surfaces, that is, along the length L. Within the box 200 spherical proteins with internal charge distribution are included (Fig. 1) having a diameter $D = 2$ nm. The hard-core interactions between proteins and the charged surfaces of implant are taken into account by means of the distance of the closest approach [13,14]. For the sake of simplicity, the hard-core interactions between proteins are neglected. The distance between the charged planar surfaces (L) is chosen to be long enough not to affect the protein distribution close to the charged surface. The canonical system was kept electroneutral in our MC simulation. The periodic boundary conditions were used in the same manner as in Ref. [15]. In each step of MC simulation, one protein was chosen randomly and randomly linearly shifted or rotated with the same probability of each kind of motion [16]. The run time was around 2×10^7 steps for the whole system of particles and the MC evaluation started after reaching the thermodynamic equilibrium of the system.

The MC simulations provided the volume charge density profile (ρ) of the protein charges in the vicinity of the Ti surface and the average order parameter $S = \langle (3 \cos^2 \Delta - 1)/2 \rangle$ within 1.25 nm wide slice (for explanation of angle Δ, please see Fig. 1). Where $S = 0$ indicates that proteins are

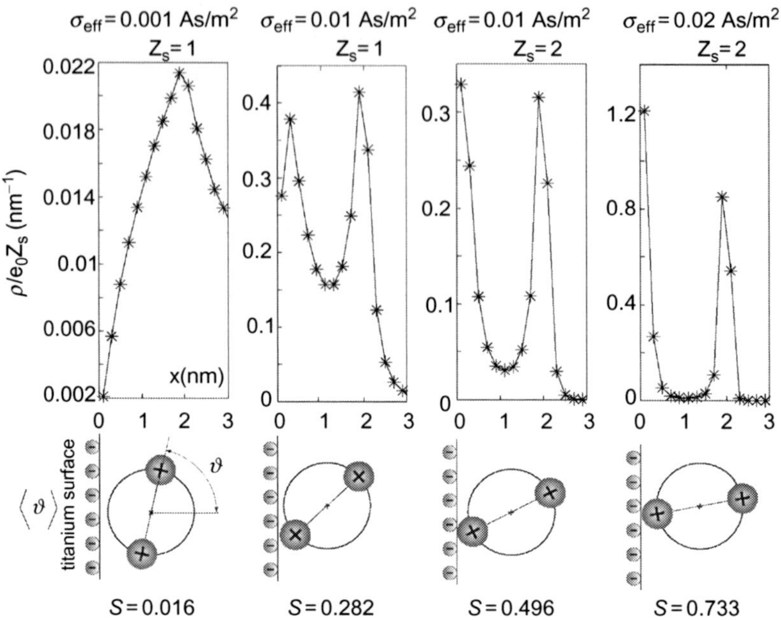

Figure 1 The calculated volume charge density of divalent macroions (ρ) as a function of distance from the charged surface is plotted for different values of the area charge density σ_{eff} and for two different valencies: $Z_s = 1$ and $Z_s = 2$. The values of the average order parameter S are computed for the slice of $x = (0,\ 1.25)$ nm in the closest vicinity of the charged surface. The average orientational angle Δ of the macroions is also shown schematically (adapted from Ref. [8]).

not oriented at all, while $S = 1$ denotes that proteins are fully oriented along the x-axis perpendicular to the charged surface.

The simulations were performed for different values of the surface charge density of the Ti surface (σ_{eff}) and different valency of the single ion within the spherical protein with internal charge distribution (Z_s). It can be seen that for high enough values of an σ_{eff} and Z_s, the concentration of the ions away from the Ti surface is zero and proteins are more oriented for higher values of the surface charge density of the charged surface and for higher values of the valency Z_s. Therefore, based on the results presented in Fig. 1, it can be concluded that for higher surface charge densities of Ti surface, the adhered proteins with an internal charge distribution would be nearly completely oriented.

In the following, we consider theoretically the solution of electrolyte (NaCl) and charged proteins in the space between planar charged Ti implant surfaces and a charged osteoblast surface with effective surface charge densities σ_{eff} and σ_{ost}, respectively.

In accordance with the conclusions of the previous part (Fig. 1), the positive charge of the tips of the bound proteins can be represented by a charged surface (with the surface charge densities σ_p) at a distance D from the Ti surface (please see inset of Fig. 2). Due to the second charge of the bound proteins, the surface charge density of the Ti surface is increased to the value $(\sigma_{\text{eff}} + \sigma_p)$.

The lipid bilayer of the osteoblasts' membrane is covered by a thick layer of charged glycoproteins with more or less homogeneous volume distribution of its negative electric charge as schematically shown in Fig. 2. In our model, we assume that spheroidal charged proteins with an internal charge distribution are embedded also in the glycoprotein layer. However, since these proteins are surrounded by a homogeneously distributed negative charge of the glycoproteins and since the charge of the lipid head groups in the outer lipid layer is negligible (see [17] and references therein), these proteins are not preferentially oriented. For the sake of simplicity in this work, the total charge distribution of negatively charged glycoprotein layer together with positively charged embedded proteins is described by an effective surface the charge density σ_{ost} (Fig. 2).

To determine the electric potential in the system, we solve the Poisson–Boltzmann equation (see, e.g., Refs. [18–20]):

$$\frac{d^2\Psi(x)}{dx^2} = \kappa^2 \sinh(\Psi(x)), \tag{1}$$

where we introduced the reduced electrostatic potential $\Psi = e_0\phi/kT$ and Debye length $l_D = \kappa^{-1} = \sqrt{\varepsilon\varepsilon_0 kT/2n_0e_0^2}$. Here, $\phi(x)$ is the electrostatic potential, e_0 is the elementary charge, kT is the thermal energy, ε is the permittivity of the medium, ε_0 is the permittivity the free space, and n_0 is the bulk number density of positively and negatively charged ions in an electrolyte solution. The axis x is perpendicular to the implant surface and points in the direction of the bulk solution. We assume the electroneutrality of the whole system and that the electrostatic field varies only in the normal direction between the two charged surfaces (i.e., in x-direction). The bulk solution (outside the space between the both charged surfaces) provides a suitable reference for the electric potential (i.e., $y = 0$). The condition of electroneutrality of the system is in agreement with the boundary conditions:

$$\frac{d\Psi}{dx}(x = 0) = -\left(\sigma_{\text{eff}} + \sigma_p\right)e_0/\varepsilon\varepsilon_0 kT, \tag{2}$$

$$\Psi(x = D_-) = \Psi(x = D_+), \tag{3}$$

$$\frac{d\Psi}{dx}(x = D_-) = \frac{d\Psi}{dx}(x = D_+) + \sigma_p e_0/\varepsilon\varepsilon_0 kT. \tag{4}$$

$$\frac{d\Psi}{dx}(x = h) = \sigma_{\text{ost}} e_0/\varepsilon\varepsilon_0 kT. \tag{5}$$

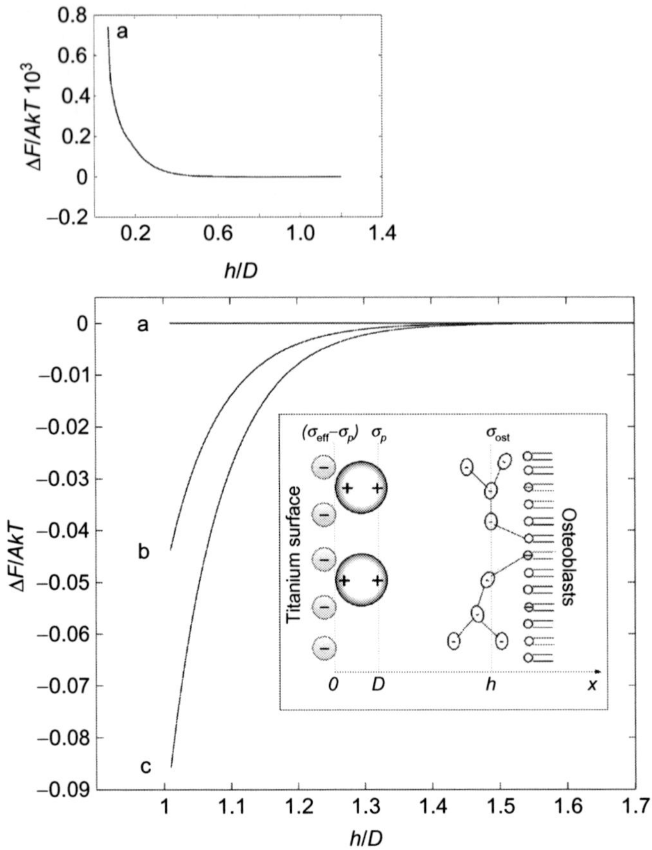

Figure 2 Free energy $\Delta F = F - F(h \rightarrow \infty)$ (in the units of AkT) as a function of h/D for different values of σ_p: 0 As m^{-2} (*a*), 0.004 As m^{-2} (*b*), and 0.008 As m^{-2} (*c*). Inset: schematic figure of a charged titanium (Ti) implant surface and a charged osteoblast's surface, interacting in an electrolyte solution. The effective surface charge density of the implant surface is σ_{eff}, while the effective surface charge density of the osteoblast surface is σ_{ost}. Due to the bound positively charged macroions with internal charge distribution, the surface charge distribution of the Ti surface is described by two surface charge densities $\sigma_{\text{eff}} + \sigma_p$ and σ_p. Upper graph: the dependence of ΔF for $\sigma_p = 0$, steric restriction due to attached proteins is not taken into account. Length of the attached proteins $D = 10$ nm. Values of other model parameters are $\sigma_{\text{eff}} = -0.4$ As m^{-2}, $\sigma_{\text{ost}} = -0.05$ As m^{-2}, salt concentration in the bulk solution $n_0/N_A = 150$ mmol l^{-1}, where N_A is the Avogadro's number.

The free energy of the system per unit area A is characterized by the energy stored in the electrostatic field (first term) and the entropic contribution of salt ions (second term) (see, e.g., Refs. [8,21]):

$$F/A = \int_0^h \left(\frac{1}{2}\varepsilon\varepsilon_0 \left(\frac{\mathrm{d}\Psi}{\mathrm{d}x}\frac{kT}{e_0} \right)^2 + kT \sum_{i=+,-} \left(n_j \ln\left(\frac{n_j}{n_0}\right) - (n_j - n_0) \right) \right) \mathrm{d}x, \tag{6}$$

where n_j are the number densities of anions ($i = -$) and cations ($i = +$) in the salt solution.

Figure 2 shows the dependency of the free energy of the system (F) on the distance h for different values of the surface charge density σ_p attributed to the bound divalent proteins. Upper graph in Fig. 2 shows the case of $\sigma_\mathrm{p} = 0$, where the attached positively charged proteins are not present in the system; therefore, the free energy F strongly increases with a decreasing distance h accordant with the strong repulsive force between Ti and osteoblast surfaces. Since for $\sigma_\mathrm{p} = 0$ there is no attached protein on the Ti surface, there is also no steric restriction for osteoblast surfaces which are therefore free to the approach to the Ti surface to the distance $h = 0$. On the other hand for $\sigma_\mathrm{p} > 0$, the proteins bound to the Ti surface prevent the closest approach of osteoblasts to Ti surface, so that $h \geq D$ (see also Fig. 2). It can be seen in Fig. 2 that for $\sigma_\mathrm{p} \geq 0$, the free energy decreases with a decreasing h until the absolute minimum of F close to $h \cong D$ is reached. It can be therefore concluded that at a large enough D and σ_p, the attached positively charged proteins with an internal charge distribution can turn the repulsive force between negatively charged Ti and osteoblast surfaces into an attractive force.

The results presented in Fig. 2 reflect the fact that two adjacent negatively charged surfaces (Ti surface and osteoblast surface) without bound positively charged proteins repel each other, while for a high enough concentration of bound positively charged proteins with an internal charge distribution, the force between two negatively charged surfaces becomes strongly attractive leading to the equilibrium distance at $h \cong D$ (Fig. 3). The

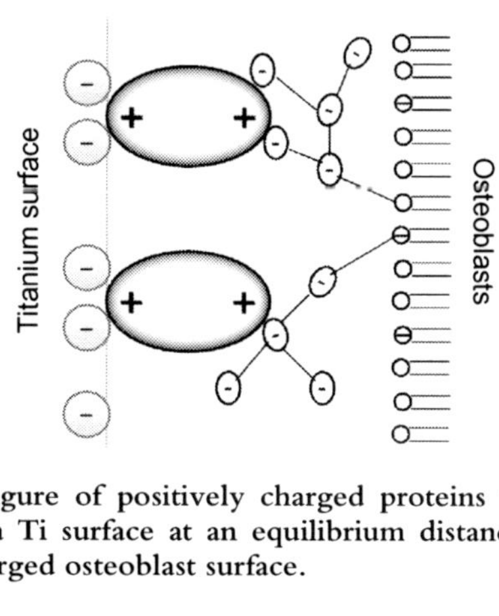

Figure 3 Schematic figure of positively charged proteins with an internal charge distribution bound to a Ti surface at an equilibrium distance between the charged implant surface and charged osteoblast surface.

origin of the attractive interactions between two negatively charged surfaces is the electrostatic attraction between the positively charged domain on the tip of the Ti surface-bound protein and negatively charged membrane osteoblast.

3. Functional Effects of a Ti Implant Surface Topography on the Osteointegration

The osteointegration on Ti implants depends on the Ti topographical characteristics. The present biomechanical chapter assumes that membrane-bound PCs with a negative intrinsic curvature reduce the effective membrane surface tension, allowing membrane growth in Ti surface grooves (Figs. 4 and 5). The binding of the osteoblasts to the Ti surface reduces the system free energy. On the other hand, the strong interactions at contact regions trap the membrane, thereby increasing the local lateral membrane tension resulting in an increase of the cell membrane free energy. In this chapter, we show that the cost of the bending energy to protrude many narrow grooves is higher than the cost to protrude a single wide groove of the same total length. We suggest that the balance between the cost of bending energy and the gain in adhesion energy may explain experimental results in which osteoblasts fail to grow in narrow grooves while growth is increased in wider grooves [10–12].

The present model suggests that the facilitated growth in wide grooves along the Ti surface is due to PCs with negative curvature-inducing domain [22–24]. Recent experimental studies have implicated Bin/amphiphysin/Rvs (BAR) and IRSp53/Missing-in-metastasis (IMD) protein domains as responsible for sensing and inducing a concave and convex curvature,

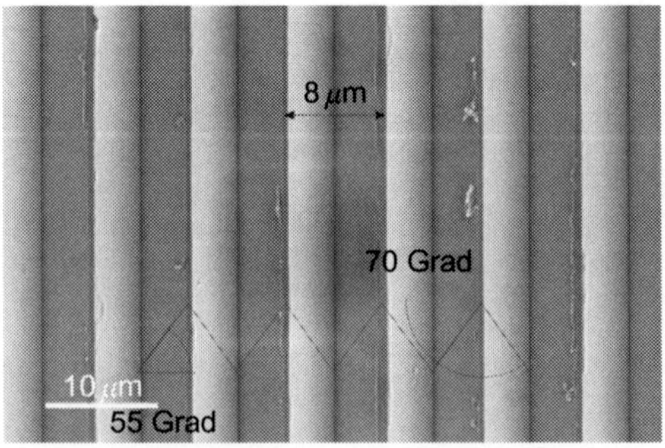

Figure 4 Electron microscope image of a Ti implant surface. Triangular profiles with V-grooves of Si-wafer coated with a thin layer of Ti. The V-grooves have an angle of 70 grad and a period of 8 μm (adapted from Ref. [8]).

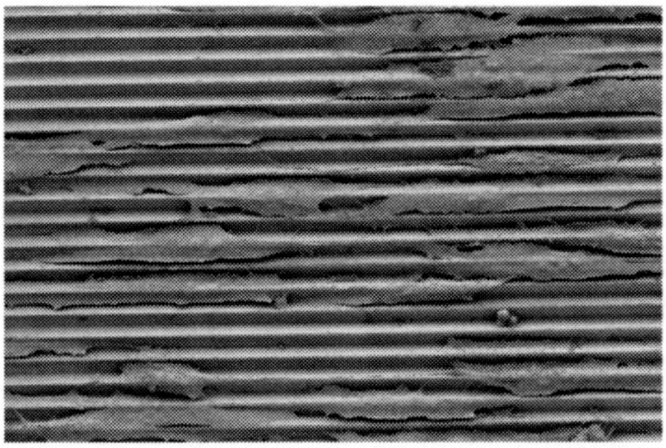

Figure 5 Electron microscope image of osteoblast cells growing along a Ti implant surface. Electron microscope image demonstrates how MG-63 osteoblast cells are growing along the V-grooves of Si-wafer coated with a thin layer of Ti (adapted from Ref. [8]).

respectively [25,26]. The BAR and IMD domains are a crescent-shaped dimer which binds preferentially to curved membranes [27] and these domains have also been shown to tubulate membranes [24,27–29]. In some of the previous theoretical studies, it was shown how such crescent-shaped proteins may induce the spontaneous initiation of tubular membrane protrusions [22,24,29–31].

In recent studies, it has been shown that adhesion molecules such as integrins connect the cell membrane to the external substrate and due to their negative binding energy, they reduce the membrane surface tension [32,33]. This binding to the external substrate sends a transmembrane signal since the adhesion molecules are also linked to the actomyosin cytoskeleton [32]. But other studies demonstrated that cells grow focal adhesions from their corner regions of a convex curvature (defined as the outward normal vector) [34]. Therefore, we hypothesize that PCs (e.g., adhesion molecules such as IRSp53) at the contact regions between the osteoblast membrane and the Ti surface have a negative intrinsic spontaneous curvature, and due to the mismatch with the membrane intrinsic curvature they initiate protrusive forces leading to osteointegration along Ti implants.

The basic equations of motion of the membrane contour and the time dependence of an nonhomogeneous lateral density of membrane PCs are derived first. A complete derivation of the system forces and fluxes is presented. We then perform a linear stability analysis for the nearly flat membrane shape.

The shapes explored in this model describe a segment of the cell outer contour, where the membrane is initially flat (Fig. 6). This geometry can also be used for a flat two-dimensional membrane under the constraint of translational symmetry. Here, we need to pin the membrane using an external force to prevent its drifting motion. This external force mimics

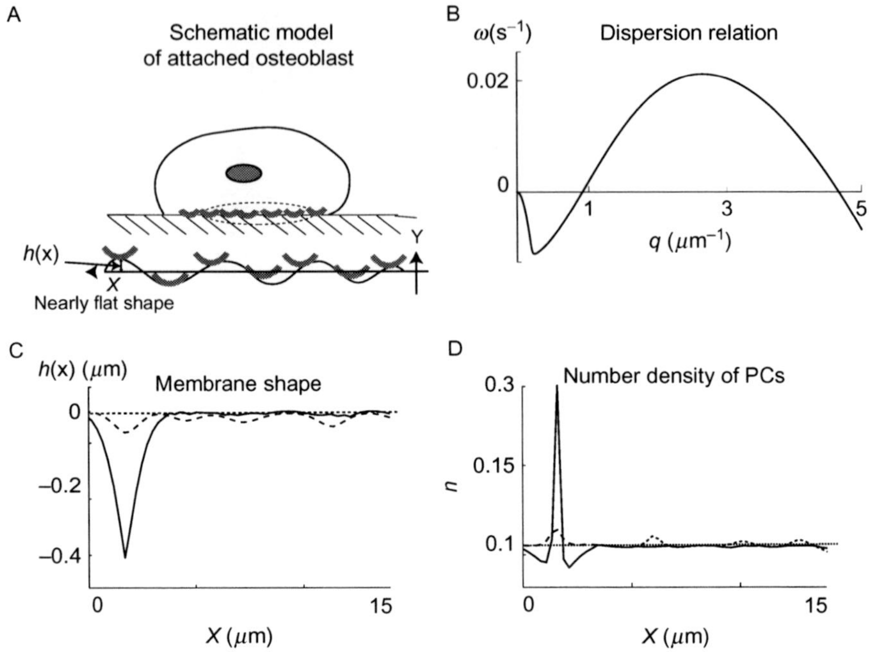

Figure 6 The dispersion relation describing the dynamical instability of membrane protrusion growth. (A) Schematic model of the flat membrane shape. The membrane shape is modeled as a two-dimensional (2D) contour in the (x, y) coordinate system describing a contour of membrane segment from an elliptical osteoblast in the XY projection. The PCs are adhesion molecules with negative intrinsic curvature, $\bar{H} = -10\,\mu m^{-1}$. (B) The wave lengths $w(q)$ is a function of different wave vectors q. Each wave vector equals $2\pi/\lambda$. The positive and negative values of $w(q)$ describe nonstable and stable regions, respectively. (C) Membrane dynamics driven by small and random perturbations in the density of PCs. At intermediate time (150 s), the membrane dynamics shows a rapid growth of four waves (dashed line) matching the fastest wave vector of 2.5 μm^{-1} of the dispersion relation. At steady state (300 s), the four membrane protrusions are coalesced into a single protrusion (solid line). (D) The dynamics of adhesion molecules (n) drives the evolution of the membrane shape. The dashed line is for intermediate time and the solid line is for steady-state time (adapted from Ref. [8]).

the effects of the restoring forces due to the adhesion of the rest of the cell to the Ti surface. In addition, it is assumed that the membrane curvature in a perpendicular direction along the contour is roughly constant (Fig. 6A), and thus, it enters our calculation as a modified membrane tension. More specifically, if the osteoblast cell is on top of the Ti surface, we only take into account a segment of the contour in contact with the Ti surface (Fig. 7A). Regarding the PCs in our model, we assume that their overall number on the membrane is conserved and that they are allowed to move along the osteoblast membrane.

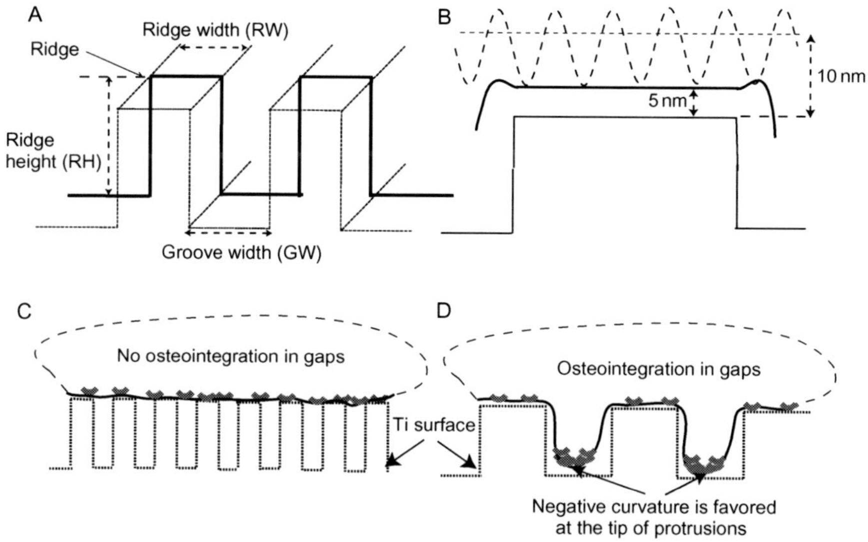

Figure 7 Schematic model of osteointegration growth along grooves of Ti surface. (A) The geometry of the Ti surface in our model. Numerical simulations test the effects of varying the Ti surface ridge width (RW), ridge height (RH), and groove width (GW). (B) The initial flattening of osteoblast membrane on top of a Ti surface ridge. Due to the strong electrostatic interactions, the membrane (solid line) is trapped crossing a threshold distance of 5 nm from the Ti surface. (C) No osteointegration is expected in narrow grooves of Ti surface. Since the density of ridges is high, large numbers of adhesion molecules (bold crescents) are trapped and the growth in grooves is not facilitated. (D) Successful osteointegration is expected in wide grooves, since the numbers of adhesion molecules (bold crescents) are sufficient to allow growth in grooves (adapted from Ref. [8]).

3.1. Derivation of Equations of Motion

We investigate the coupling between the adhesion of osteoblast cells and the membrane curvature. The first hypothesis is that the negative binding energy due to the adhesion of proteins on the outer surface of osteoblast cells could initiate the growth of a membrane protrusion. The second hypothesis is that the coupling between the density of proteins and the membrane curvature would lead to a positive feedback loop facilitating the growth of a membrane protrusion. We give below the free energy expression used in the model, from which we derive the equations of motion of the membrane contour and density distribution.

The continuum free energy in our model is [31,35]

$$F = \int \left(\frac{1}{2}\kappa(H - \bar{H}n)^2 + (\sigma - \alpha n) + \gamma h^2 + k_B T n_s n(\ln(n) - 1) \right) ds, \quad (7)$$

where the first term gives the bending energy [35] due to the mismatch between the membrane curvature and the spontaneous curvature of the PC. The second term describes the negative contribution of PCs (e.g., adhesion molecules) to the membrane surface tension. The third term gives the energy due to the force of the cytoskeleton inside osteoblasts. The fourth term gives the entropic contribution due to the lateral thermal motion of the PC in the membrane in the limit of small n [22]. Here, κ is the membrane bending constant [35], H is the local mean membrane curvature, $\bar{H}$ is the intrinsic curvature of the PC, n is the area fraction of PCs, n_s is the saturation density of PCs on the membrane, σ is the membrane surface tension, α is proportionality constant describing the effective interaction between the PCs and the external substrate, $h = h(x)$ describes the magnitudes of small deformations from the flat membrane, γ is a restoring spring constant, and $\mathrm{d}s = d \cdot \mathrm{d}l$ is an the element of membrane area, where d is the dimension of membrane perpendicular to the contour and $\mathrm{d}l$ is a line element along the contour. The local mean curvature is equal to $H = (\partial^2 y/\partial s^2)(\partial x/\partial s) - (\partial^2 x/\partial s^2)(\partial y/\partial s)$. The discretization, $\mathrm{d}s$, equals $\sqrt{(\partial x/\partial s)^2 + (\partial y/\partial s)^2}$. Equation (7) is a one-dimensional version of the more general expressions derived in Refs. [36,37]. These expressions recover the familiar form for small undulations of a flat membrane in the Monge approximation.

3.2. Membrane Forces

We next derive the forces on the membrane, by treating it as a "one-dimensional membrane," that is, a thin strip of a width w, with a bending modulus and tension coefficient. The free energy of this membrane is given in Eq. (7) and is used to derive the local restoring forces by the usual variation method. The variation of the free energy (Eq. 7) is with respect to the membrane coordinate [38,39] and PC concentration [38].

The derivation of the first term, that is, the Helfrich part in the free energy ($1/2\kappa H^2$), was already derived in a previous study [38]. The following is the derivation of the second term $Hn\dot{s}$. The third term obtained after opening the first parenthesis in the free energy is $\bar{H}^2 n^2$. Since it is not dependent, on x or y, it is treated as a constant times $\dot{s}$. The other terms in the free energy may also be treated as a constant times $\dot{s}$. The force derived from the variation of a constant times $\dot{s}$ is the constant itself times H [38].

Since the overall contour length is not constant in our system, the variation of the coordinates has to be taken with respect to their absolute index u along the contour, which is constant. In these terms, the curvature H appearing in the free energy is written as a (standard differential geometry)

$$H = \frac{\dot{x}\ddot{y} - \ddot{x}\dot{y}}{\sqrt{\dot{x}^2 + \dot{y}^2}^3}, \tag{8}$$

where the $\cdot$ symbol denotes differentiation with respect to the index of the point along the contour. The ds in the free energy (Eq. 7) is replaced with $\sqrt{\dot{x}^2 + \dot{y}^2}\mathrm{d}u$, such that $\mathrm{d}s/\mathrm{d}u = \dot{s} = \sqrt{\dot{x}^2 + \dot{y}^2}$. The variation of the free energy gives the forces, for example, in the x-direction:

$$F_x = -\frac{\delta \mathcal{F}}{\delta x} = \frac{\mathrm{d}}{\mathrm{d}u}\frac{\partial \mathcal{F}}{\partial \dot{x}} - \frac{\mathrm{d}^2}{\mathrm{d}u^2}\frac{\partial \mathcal{F}}{\partial \ddot{x}}. \tag{9}$$

The resulting equation of motion from this variation gives very long expressions that are not amenable to an easy analysis, although they can be used for the numerical simulations. To arrive at simpler expressions, we will develop the terms in Eq. (9) and simplify at the end by assuming that the arc-length separation between the nodes along the contour are all the same. As explained earlier, the variation of the Helfrich part was already derived in a previous study. Herein, we derive the contributions of the term $\bar{F} = Hn\dot{s}$? to the forces in our system. For the sake of simplicity, we leave out the constants $\bar{H}$ and $1/2\kappa$. The first term on the RHS of Eq. (9) is

$$\frac{\partial \bar{\mathcal{F}}}{\partial \dot{x}} = \frac{\ddot{y}n}{\dot{s}^2} - 2Hn\frac{\dot{x}}{\dot{s}}, \tag{10}$$

whereby using the following identities, $\dot{x}/\dot{s} = x'$, $\ddot{y} = \ddot{s}y' + \dot{s}^2x'H$, and $\ddot{s} = 0$ ($\dot{s}$ is independent of u), we obtain

$$\frac{\partial \bar{\mathcal{F}}}{\partial \dot{x}} = -Hnx'. \tag{11}$$

The first term of Eq. (9) is

$$\begin{aligned} \frac{\mathrm{d}}{\mathrm{d}u}\frac{\partial \bar{\mathcal{F}}}{\partial \dot{x}} &= \dot{s}\frac{\mathrm{d}}{\mathrm{d}s}(-Hnx') \\ &= \dot{s}\left(-H'nx' - Hn'x' - Hnx''\right). \end{aligned} \tag{12}$$

The second term on the RHS of Eq. (9) is

$$\begin{aligned} \frac{\partial \bar{\mathcal{F}}}{\partial \ddot{x}} &= \frac{\partial}{\partial \ddot{x}}(Hn\dot{s}) \\ &= \frac{-y'n}{\dot{s}}, \end{aligned} \tag{13}$$

where we used the equality $\dot{y}/\dot{s} = y'$. By taking the second derivative, we get

$$\begin{aligned} \frac{\mathrm{d}^2}{\mathrm{d}u^2}\frac{\partial \mathcal{F}}{\partial \ddot{x}} &= \dot{s}^2 \frac{\partial^2}{\partial s^2}\left(\frac{-y'n}{\dot{s}}\right) \\ &= \dot{s}\frac{\partial}{\partial s}\left(-y''n - y'n'\right) \\ &= \dot{s}\left(-y'''n - 2y''n' - y'n''\right). \end{aligned} \tag{14}$$

Adding the two terms on the RHS of Eq. (9), the force along the x-direction is

$$F_x = \dot{s}\left(\left(-H'nx' - Hn'x' - Hnx''\right) - \left(-y'''n - 2y''n' - y'n''\right)\right). \tag{15}$$

In a similar manner, the derivation of the force along the y-direction gives

$$F_y = \dot{s}\left(\left(-H'ny' - Hn'y' - Hny''\right) - \left(x'''n + 2x''n' + x'n''\right)\right). \tag{16}$$

The calculations above are for forces per unit u. Since we need forces per unit length, we divide by $\dot{s}$ to obtain the following force. The normal force acting on the membrane is

$$\begin{aligned} F_n &= \vec{F}\cdot\hat{n} = -y'F_x + x'F_y \\ &= -y'\left(\left(-H'nx' - Hn'x' - Hnx''\right) - \left(-y'''n - 2y''n' - y'n''\right)\right) \\ &\quad + x'\left(\left(-H'ny' - Hn'y' - Hny''\right) - \left(x'''n + 2x''n' + x'n''\right)\right) \\ &= -n'', \end{aligned} \tag{17}$$

where we used the following identities, $H = x'y'' - y'x''$, $x'^2 + y'^2 = 1$, $2(x'x'' + y'y'') = \partial\,(x'^2 + y'^2)/\partial\, s = 0$, and the second derivative of $x'x'' + y'y''$ which gives another identity $x'x''' + y'y''' = -H^2$, where $H^2 = x''^2 + y''^2$.

To take into account the drag due to viscous forces, we assume for simplicity only local friction forces [31,38], with coefficient ξ. The equation of motion is given by

$$\xi\frac{\partial\vec{r}}{\partial t}\cdot\vec{n} = -\frac{\delta F(s,t)}{\delta h}, \tag{18}$$

where $\vec{r}$ is the (x, y) coordinates, s is the contour length, t is the time, $\vec{n}$ is the normal direction, and $-\delta F(s, t)/\delta h$ is the force derived from the variation. Since $F_{\mathrm{n}} = -\delta F(s, t)/\delta h$, the added contribution to the force due to the curvature mismatch term is $\kappa \bar{H} \nabla^2 n$. Note that the force $\delta F(s, t)/\delta h$ equals the membrane shape velocity times friction coefficient, but opposite in direction. Since the relative change in the y (i.e., vertical)-direction is considerably greater than the change along the x (i.e., horizontal)-direction, we only consider the changes along the y-direction. This simplification is a variation of the Monge implementation, where the forces are applied along the vertical direction. In our case, it is satisfactory since we limit the change along the y-direction not to exceed a maximum ridge height (< 0.2 μm). Due to these reasons, the results will remain qualitatively the same if the nonlinear term $\frac{1}{2}H^3$ in the following equation for $F_{\mathrm{curvature}}$ is neglected.

In the free energy, we use a nonzero spring term to prevent a drift due to the activity of an adhesion. The variation of the free energy is projected to give the forces normal to the membrane contour [38,39]. We now list the forces derived from the variation of the free energy (Eq. 7) [8,38,39]:

$$F_{\mathrm{curvature}} = \kappa \left(-\nabla^2 H + \bar{H} \nabla^2 n + \frac{1}{2} n^2 \bar{H}^2 H + \frac{1}{2} H^3 \right), \tag{19}$$

$$F_{\mathrm{tension}} = (\sigma - \alpha n) H, \tag{20}$$

$$F_{\mathrm{spring}} = -2\gamma y, \tag{21}$$

$$F_{\mathrm{entropy}} = kTn_{\mathrm{s}}(n \ln n - 1) H, \tag{22}$$

where $F_{\mathrm{curvature}}$ is the force due to the curvature energy mismatch between the membrane curvature and the spontaneous curvature of the PCs, F_{tension} is the membrane tension force, F_{spring} is the spring restoring force, and F_{entropy} arises from the entropy of the PCs in the membrane, which acts to expand the length of the contour. To control the growth in the cell contour length, we made use of a nonlinear form for the effective membrane tension:

$$\sigma = \sigma_0 \exp[\beta(L_{\mathrm{t}} - L_{\mathrm{init}})], \tag{23}$$

where L_{t} is the contour length, L_{init} is the initial contour length, and β is the factor that determines the length scale at which the nonlinear growth in the tension sets in. Finally, the strong electrostatic interactions trap the membrane and prevent its further evolution throughout the simulation. This trapping occurs below a threshold distance of 5 nm.

3.3. The Intramembrane Protein Fluxes

The intramembrane fluxes of proteins are derived from the free energy (Eq. 7). The dynamics of the PC density, using the following conservation equation, is as follows:

$$\frac{\partial n}{\partial t} = -\nabla \cdot \overrightarrow{J} = \frac{\Lambda}{n_s} \nabla \left(n \nabla \frac{\delta F}{\delta n} \right) - \frac{n}{\mathrm{d}l} \frac{\partial \mathrm{d}l}{\mathrm{d}t}, \tag{24}$$

where Λ is the mobility of filaments and $\overrightarrow{J}$ is the total current of PC on the membrane. The total flux J is

$$J = -\frac{\Lambda}{n_s} \nabla \left(n \nabla \frac{\delta F}{\delta n} \right). \tag{25}$$

The derivation of the free energy with respect to n gives

$$\frac{\delta F}{\delta n} = \kappa (H - \bar{H} n)(-\bar{H}) - \alpha. \tag{26}$$

The differentiation with respect to the contour length s is

$$\nabla \left(\frac{\delta F}{\delta n} \right) = \kappa (-\bar{H}) \nabla H + \kappa \bar{H}^2 \nabla n. \tag{27}$$

The resulted fluxes are

$$J = \frac{\kappa \Lambda \bar{H}}{n_s} n \nabla H - \frac{\kappa \Lambda \bar{H}^2}{n_s} n \nabla n. \tag{28}$$

where $(\kappa \Lambda \bar{H}/n_s) n \nabla H$ is the "attraction" flux resulting from the interaction between the proteins through the membrane curvature, and $-\left(\kappa \Lambda \bar{H}^2/n_s\right) n \nabla n$ is the "dispersion" flux due to the membrane resistance to protein aggregation due to their membrane bending effects. Another flux that is not derived from the free energy is the flux due to usual thermal diffusion $-D \nabla n$ flux, where D is the diffusion constant. The last term in Eq. (24) arises from the covariant derivative of the density with time on a contour whose length evolves with time. In this term, dl is simply the line element. This term ensures that the total number of PC is conserved as the contour length changes.

3.4. Linear Stability Analysis

We next performed a linear stability analysis of the flat membrane model, as was previously done in Ref. [31]. The contour is allowed to evolve only along one direction and we label the amplitude of the membrane fluctuation as $h(x)$, where x is the coordinate along the initial contour length. The curvature is linearized to be $H \simeq \nabla^2 h$ and the length element of the contour dl is given by d$l = 1 + (\nabla h)^2/2$. Using this linearization, and linearizing the equations of motion (Eqs. 18 and 24), we then apply Fourier transform to get a 2×2 matrix whose eigenvalues give the dynamic evolution of small fluctuations from the equilibrium flat state. Both eigenvalues are real, and one of them is always negative and therefore represents only damped modes. The second solution can become positive in a range of wave vectors, representing unstable modes that grow with time. The following is the list of parameter values incorporated in the dispersion relation and numerical simulations: $\xi = 125\ \mathrm{g\ s^{-1}}$, $D_1 = 0.002\ \mu\mathrm{m}^2\ \mathrm{s}^{-1}$, $\alpha = 0.013\ \mathrm{g\ s^{-2}}$, $\gamma = 0.00004\ \mathrm{g\ s^{-2}}$, $n_s = 10\ \mu\mathrm{m}^{-2}$, $\kappa = 100kT$, $\bar{H} = -10\,\mu\mathrm{m}^{-1}$, and $\sigma = 0.001\ \mathrm{g\,s^{-2}}$ [40]. According to a typical diameter of an osteoblast cell, the initial length of the nearly flat-modeled membrane is 15 μm. The total number of molecules remains fixed throughout the simulation. The total number of molecules is the total length of the modeled membrane (15 μm) times the average fractional density (0.1) divided by the length of a typical PC (10 nm) to give 150 molecules. The initial conditions of PCs are a uniform distribution of density 0.1 with small (1% of the maximum amplitude) random noise. The flat membrane shape is initially positioned 10 nm above the Ti ridge height (Fig. 7).

There are two sources of dynamical instability: the first source originates from the mismatch between the spontaneous curvature of PCs and the membrane curvature and the second source is due to adhesion forces which effectively reduce the membrane tension. The result of this dynamical instability is a dispersion relation of the eigenvalues $w(q)$ obtained from the aforementioned matrix. This dispersion relation has a nonstable region engulfed by two stable regions. In a nonstable region, a small perturbation ($<1\%$) is sufficient to initiate exponential growth, while a small perturbation in a stable region will decay exponentially. When the initial conditions are in the nonstable region, the growing membrane protrusions coalesce into a single pointed peak. At this region, the filament density is considerably greater than its initial density. In addition, the distribution of PCs matches the distribution of convex regions along the nearly flat membrane shape (Fig. 6). This result is due to the attraction of PCs to convex regions (with negative curvature) that match their intrinsic spontaneous curvature of $-10\ \mu\mathrm{m}^{-1}$.

3.5. Numerical Simulations

Previous experimental studies have shown that osteoblasts do not grow in narrow grooves of the Ti implant surface, while successful growth is observed in wide grooves. The present model evaluates the functional effects on osteointegration by varying the three morphologic parameters of Ti surface shape, that is, the groove width, the ridge height, and the ridge width (Figs. 7 and 8). In addition, we will lower the energy cost of bending

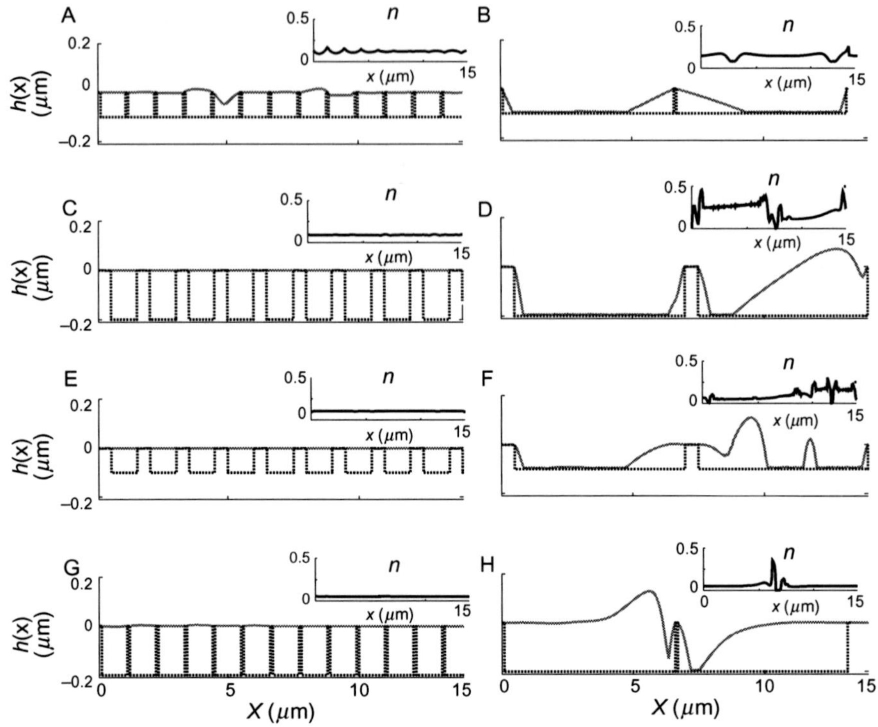

Figure 8 The effects of varying the ridge width, ridge height, and groove width on osteointegration. Each panel shows the osteoblast shape growth ($h(x)$; solid line) over the Ti surface (dotted line). Each inset shows the distribution of adhesion molecules (n; intrinsic curvature, $\bar{H} = -10\ \mu m^{-1}$) projected along the x-dimension of the Ti surface. The steady-state time of the membrane shape ranges between 300 and 1000 s. The analysis of the Ti surface shape includes changes in the groove width, ridge height, and ridge width. The osteoblast growth is determined for a groove width of 1 μm (A, C, E, G) versus 6.5 μm (B, D, F, H), a ridge height of 0.1 μm (A, B, E, F) versus 0.2 μm (C, D, G, H), and a ridge width of 0.5 μm (A, B, G, H) versus 2 μm (C–F). Due to strong electrostatic interactions between the osteoblast shape and the Ti surface, we take into account an adhesion zone of 5 nm strip thickness above the Ti surface. Note the large growth seen in grooves of 6.5 μm versus little growth seen along a groove width of 1 μm. In these simulations, the bending constant, κ, equals $100kT$ (adapted from Ref. [8]).

by reducing the bending modulus κ, demonstrating that the osteoblast can then grow in narrow grooves. The dynamics of membrane shape includes electrostatic interactions (at small length scale <5 nm) and dynamic instability due to adhesive forces (at large length scale >5 nm). The strong electrostatic interactions pin the membrane locally and prevent its further evolution with time. In fact, the membrane coordinates that trespass this adhesion zone at <5 nm distance from the Ti surface will remain fixed to the end of the simulation (Fig. 7). Note that this adhesion zone is expected to increase the stability in the system.

On the other hand, the growth of membrane protrusions between the Ti surface grooves is driven by instability due to a large adhesive strength, which effectively reduces the membrane tension (Fig. 8). Our analysis reveals that the morphologic parameter that has the largest effect on osteoblast integration is the groove width. In particular, Fig. 8 demonstrates that over a groove width of 6.5 μm, the osteointegration is greater than in narrow grooves of 1 μm. The ridge width also plays a crucial role in determining the amounts of strong electrostatic interactions which pin the membrane and increase the membrane stiffness. In addition, less osteointegration is observed for a ridge width of 0.5 μm (second and third rows) in comparison to a ridge width of 0.1 μm (first and fourth rows). As predicted from the theory, we find that the distribution of PCs follows the membrane shape dynamics, where at regions of osteointegration there is a large aggregation of adhesion molecules. To determine the role of the bending constant, we compared the osteoblast protrusion in 3 μm grooves for different values of κ (not shown). Results of these simulations reveal that the osteoblast protrusion in narrow groove was considerable only for the lower bending constant, demonstrating the important role of bending energy to the success of osteointegration.

To have a better understanding of the underlying biophysical mechanisms allowing osteointegration only in wide grooves, we determined the contributions of fluxes due to PC spontaneous curvature to osteointegration. Numerical simulations revealed that the distribution of PCs followed the distribution of membrane curvature. At protrusive regions, the density of PCs as well as the fluxes due to membrane curvature were high. On the other hand, since the membrane regions above the ridge surface were nearly flat, the fluxes due to membrane curvature were diminished. While the relative contributions of fluxes due to diffusion were minor, the PC density at these regions was close to the initial value, n_0. Therefore, the ridge region poses an energy barrier over which the PCs cannot cross through, and the remaining number of PCs in the narrow grooves was not sufficient for successful osteointegration.

4. Conclusions

In accordance with previous suggestions [6,9,12], we have shown that positively charged proteins attached to the negatively charged implant surface can mediate the adhesion of negatively charged osteoblasts to negatively charged Ti surface. As shown in Fig. 1, the protein adsorption to the implant surface may be driven by electrostatic interactions between the positive sites of the protein and the negative implant surface [41]. Another possible or additional explanation of the adsorption of charged proteins on the Ti implant surface may include also the monovalent and divalent salt bridges [42,43]. We also investigated how the mediated protein adsorption and strong electrostatic interactions with the Ti implant may affect the dynamics of osteointegration at a greater distance from the implant.

The adhesion of osteoblast-like cells (i.e., bone cells) to a Ti surface implant is a dynamic process driven by the interaction with the extracellular matrix as well as intracellular mechanisms. It has been shown that the adhesion to the extracellular matrix mediates the transport of curvature-inducing adhesion molecules to the membrane, thereby relieving the surface membrane tension and allowing the membrane protrusive growth [31,38].

In addition, the membrane protrusive growth is also mediated by a positive feedback between the aggregation of curvature-generating molecules (e.g., IRSp53 and fibronectin) and the membrane curvature. While the aforementioned mechanisms increase the instability in our system, the strong adhesion to the Ti surface due to electrostatic interactions will increase the system stability as well as the membrane surface tension. In recent theoretical studies, the use of an external actin nucleation force was coupled to the density of PCs [31,38]. Since the nucleation of actin was mainly at membrane regions of high PC density, this mechanism can also be applied to explain the growth of membrane protrusions in between Ti surface ridges.

In the first set of presented biomechanical simulations, we investigate the initial membrane protrusive growth over a ridge-like Ti surface pattern in which the following three morphologic features are varied: the ridge height, the ridge width, and the groove width. The observed flattening of the growing membrane tips against the Ti surface diminishes the fluxes of PCs and the lateral surface membrane tension is increased. We also find that the groove width has the largest impact on osteointegration, such that a wide groove facilitates the membrane ingrowth between ridges. On the other hand, a high ratio of the ridge width over the groove width will increase the size of trapped and flattened membrane regions, thereby reducing the amounts of available PCs needed for the successful osteointegration.

To isolate the contribution of the bending modulus, simulation results reveal that by lowering of the bending modulus κ, the osteointegration in relatively narrow grooves (3 μm) was increased in comparison to the reference bending modulus. This shows that in a plausible range of κ, the results of simulations may change considerably, demonstrating the important contribution of the bending energy to the osteointegration.

The main result of the biomechanical part of our study is that the decreased width of the grooves results in a decreased adhesion and ingrowth between the grooves, which is supported by various experimental studies

[10–12]. In Ref. [12], it has been also shown that decreasing the width of nanorough regions (of dimension from 80 and 48 μm to 22 μm) resulted in significant reductions in the number of osteoblast cells adhering to the structured surface. One explanation is that due to their large cell size (20–30 μm in length), osteoblast cells are unable to fit their entire length into a narrow groove without a membrane bending. Results in the present chapter reveal that the observed reduction in the cell adhesion could be due to a high bending energy, which reduces the osteointegration in between narrow grooves.

To summarize, the present chapter demonstrated how positively charged proteins mediate the adhesion of negatively charged osteoblasts to a negatively charged implant surface. In addition to this mathematical analysis, a dynamical model investigated the contribution of the bending energy of an initially nearly flat membrane to the ingrowth of filopodia-like protrusion between Ti surface ridges. We showed that the strong electrostatic binding of the cell membrane to the Ti surface effectively sequesters the PCs from driving the membrane ingrowth away from the bounded regions. In addition, the success of osteointegration between grooves depended on the groove width. In particular, the bending energy involved in folding the cell membrane along a narrow groove was too high to be compensated by the reduced membrane tension mediated by the adhesion of the osteoblast cell to the extracellular matrix. Finally, it remains to be tested how the bending of the entire osteoblast cell is affected by the implant surface pattern, and how it will affect the cell alignment, cell orientation, and cell motility.

REFERENCES

[1] K. Cai, M. Frant, J. Bossert, G. Hildebrand, K. Liefeith, K.D. Jandt, Surface functionalized titanium thin films: zeta-potential, protein adsorption and cell proliferation, Colloids Surf. B Biointerfaces 50 (2006) 1–8.

[2] S. Roessler, R. Zimmermann, D. Scharnweber, C. Werner, H. Worch, Characterization of oxide layers on Ti6Al4V and titanium by streaming potential and streaming current measurements, Colloids Surf. B Biointerfaces 26 (2002) 387–395.

[3] H.J. Butt, K. Graf, M. Kappl, Physics and Chemistry of Interfaces, first ed. Wiley-VCH Verlag, Weinheim, 2006.

[4] T.K. Monsees, K. Barth, S. Tippelt, K. Heidel, A. Gorbunov, W. Pompe, et al. Surface patterning on adhesion, differentiation, and orientation of osteoblast-like cells, Cells Tissues Organs 180 (2005) 81–95.

[5] F.F. Walboomers, J.A. Jansen, Cell and tissue behaviour on micro-grooved surface, Odontology 89 (2001) 2–11.

[6] R. Smeets, A. Kolk, M. Gerressen, O. Driemel, O. Maciejewski, B. Hermanns-Sachweh, et al. A new biphasic osteoinductive calcium composite material with a negative zeta potential for bone augmentation, Head Face Med. 5 (2009) 1310.1186/1746-160X-5-13.

[7] N.C. Teng, S. Nakamura, Y. Takagi, Y. Yamashita, M. Ohgaki, K. Yamashita, A new approach to enhancement of bone polarized hydroxyapatite, J. Dent. Res. 80 (2000) 1925–1929.

[8] D. Kabaso, E. Gongadze, S. Perutková, V. Kralj-Iglič, C. Matschegewski, U. Beck, et al. Mechanics and electrostatics of the interactions between osteoblasts and titanium surface, Comput. Methods Biomech. Biomed. Eng. 2011(in press).

[9] I.O. Smith, M.J. Baumann, L.R. McCabe, Electrostatic interactions as a predictor for osteoblast attachment to biomaterials, J. Biomed. Mater. Res. A 70 (2004) 436–441.

[10] E. Lamers, X.F. Walboomers, M. Domanski, J. te Riet, F.C. van Delft, R. Luttge, et al. The influence of nanoscale grooved substrates on osteoblast behavior and extracellular matrix deposition, Biomaterials 31 (2010) 3307–3316.

[11] C. Matschegewski, S. Staehlke, R. Loeffler, R. Lange, F. Chai, D.P. Kern, et al. Cell architecture cell function dependencies on titanium arrays with regular geometry, Biomaterials 31 (2010) 5729–5740.

[12] S. Puckett, R. Pareta, T.J. Webster, Nano rough micron patterned titanium for directing osteoblast morphology and adhesion, Int. J. Nanomed. 3 (2008) 229–241.

[13] S. Perutková, M. Frank, K. Bohinc, G. Bobojevič, B. Rozman, V. Kralj-Iglič, et al. Interaction between equally charged membrane surfaces mediated by positively and negatively charged macroions, J. Membr. Biol. 236 (2010) 43–53.

[14] J. Urbanija, K. Bohinc, A. Bellen, S. Maset, A. Iglic, V. Kralj-Iglic, et al. Attraction between negatively charged surfaces mediated by spherical counterions with quadrupolar charge distribution, J. Chem. Phys. 129 (2008) 105101.

[15] A.G. Moreira, R.R. Netz, Simulations of counterions at charged plates, Eur. Phys. J. E 8 (2002) 33–58.

[16] D. Frenkel, B. Smith, Understanding Molecular Simulation from Algorithms to Applications, Academic Press, London, 2002.

[17] A. Iglič, M. Brumen, S. Svetina, Determination of inner surface potential of erythrocyte membrane, Bioelectrochem. Bioenerg. 43 (1997) 97–103.

[18] G. Cevc, Membrane electrostatics, Biochim. Biophys. Acta 1031 (1990) 311–382.

[19] J.N. Israelachvili, Intermolecular and Surface Forces, Academic Press, San Diego, CA, 1995.

[20] S.A. Safran, Statistical Thermodynamics of Surfaces, Interfaces, and Membranes, Westview Press, Boulder, CO, 2003.

[21] K. Bohinc, T. Slivnik, A. Iglič, V. Kralj-Iglič, Membrane electrostatics—a statistical mechanical approach to the functional density theory of electric double layer, in: A. Leitmannova Liu (Ed.), Advances in Planar Lipid Bilayers and Liposomes (APLBL), vol. 8. Elsevier, Amsterdam, 2008.

[22] A. Iglič, H. Hägerstrand, P. Veranič, A. Plemenitaš, V. Kralj-Iglič, Curvature induced accumulation of anisotropic membrane components and raft formation in cylindrical membrane protrusions, J. Theor. Biol. 240 (2006) 368–373.

[23] A. Iglič, M. Lokar, B. Babnik, T. Slivnik, P. Veranič, H. Hägerstrand, et al. Possible role of flexible red blood cell membrane nanodomains in the growth and stability of membrane nanotubes, Blood Cells Mol. Dis. 39 (2007) 14–23.

[24] Y.J. Zimmerberg, M.M. Kozlov, How proteins produce cellular curvature, Nat. Rev. Mol. Cell Biol. 7 (2006) 9–19.

[25] P.K. Mattila, A. Pykalainen, J. Saarikangas, V.O. Paavilainen, H. Vihinen, E. Jokitalo, et al. Missing-in-metastasis and IRSp53 deform PI(4,5)P2-rich membranes by an inverse bar domain-like mechanism, J. Cell Biol. 176 (2007) 953–964.

[26] B.J. Peter, H.M. Kent, I.G. Mills, Y. Vallis, P.J. Butler, P.R. Evans, et al. Bar domains as sensors of membrane curvature: the amphiphysin bar structure, Science 303 (2004) 495–499.

[27] R. Shlomovitz, N.S. Gov, Physical model of contractile ring initiation in dividing cells, Biophys. J. 94(16) (2008) 1155–1168.
[28] A. Iglič, T. Slivnik, V. Kralj-Iglič, Elastic properties of biological membranes influenced by attached proteins, J. Biomech. 40 (2007) 2492–2500.
[29] S. Perutková, V. Kralj-Iglič, M. Frank, A. Iglič, Mechanical stability of membrane nanotubular protrusions influenced by attachment of flexible rod-like protein, J. Biomech. 43 (2010) 1612–1617.
[30] B. Sorre, A. Callan-Jones, J.B. Manneville, P. Nassoy, J.F. Joanny, J. Prost, et al. Curvature-driven lipid sorting needs proximity to a demixing point and is aided by proteins, Proc. Natl. Acad. Sci. USA 106 (2009) 5622–5626.
[31] A. Veksler, N.S. Gov, Phase transitions of the coupled membrane-cytoskeleton modify cellular shape, Biophys. J. 93(11) (2007) 3798–3810.
[32] D.E. Discher, P. Janmey, Y.L. Wang, Tissue cells feel and respond to the stiffness of their substrate, Science 310 (2005) 5751.
[33] A. Engler, L. Bacakova, C. Newman, A. Hategan, M. Griffin, D. Discher, Substrate compliance versus ligand density in cell on gel responses, Biophys. J. 86 (2004) 617–628.
[34] A. Brock, E. Chang, C.C. Ho, P. LeDuc, X. Jiang, G.M. Whitesides, et al. Geometric determinants of directional cell motility revealed using microcontact printing, Langmuir 19 (2003) 1611–1617.
[35] W. Helfrich, Elastic properties of lipid bilayers: theory and possible experiments, Z. Naturforsch. C 28 (1973) 693–703.
[36] W. Cai, T.C. Lubensky, Covariant hydrodynamics of fluid membranes, Phys. Rev. Lett. 73 (1994) 1186–1189.
[37] Z.C. Tu, Z.C. Ou-Yang, Lipid membranes with free edges, Phys. Rev. E 68 (2003) 061915.
[38] D. Kabaso, R. Shlomovitz, T. Auth, V.L. Lew, N.S. Gov, Curling and local shape changes of red blood cell membranes driven by cytoskeletal reorganization, Biophys. J. 99 (2010) 808.
[39] S. Mark, R. Shlomovitz, N.S. Gov, M. Poujade, E. Grasland-Mongrain, P. Silberzan, Physical model of the dynamic instability in an expanding cell culture, Biophys. J. 98 (2010) 361–370.
[40] R. Simson, E. Wallraff, J. Faix, J. Niewohner, G. Gerisch, E. Sackmann, Membrane bending modulus and adhesion energy of wild-type and mutant cells of dictyostelium lacking talin or cortexillins, Biophys. J. 74 (1998) 514–522.
[41] D.D. MacDonald, B. Markovic, A.L. Boskey, P. Somasundaran, Physico-chemical properties of human plasma fibronectin binding to well characterized titanium dioxide, Colloids Surf. B Biointerfaces 11 (1998) 131–139.
[42] J.E. Ellingsen, A study on the mechanism of protein adsorption to TiO_2, Biomaterials 12 (1991) 593–596.
[43] Y. Yang, R. Glover, J.L. Ong, Fibronectin adsorption on titanium surfaces and its effect on osteoblast precursor cell attachment, Colloids Surf. B Biointerfaces 30 (2003) 291–297.

Subject Index

Lightning Source UK Ltd.
Milton Keynes UK

176272UK00004B/29/P

9 780123 877215